MANAGING DATA USING EXCEL

OTHER PELAGIC PUBLISHING TITLES BY THIS AUTHOR

Mark Gardener (2012) *Statistics for Ecologists Using R and Excel: Data Collection, Exploration, Analysis and Presentation.* Exeter: Pelagic Publishing, UK.

Mark Gardener (2014) *Community Ecology: Analytical Methods Using R and Excel.* Exeter: Pelagic Publishing, UK.

Managing Data Using Excel®

Organizing, summarizing and visualizing scientific data

Mark Gardener

RESEARCH SKILLS SERIES

Pelagic Publishing | www.pelagicpublishing.com

Published by Pelagic Publishing
www.pelagicpublishing.com
PO Box 725, Exeter EX1 9QU, UK

Managing Data Using Excel: Organizing, Summarizing
and Visualizing Scientific Data

ISBN 978-1-78427-007-0 (Pbk)
ISBN 978-1-78427-008-7 (Hbk)
ISBN 978-1-78427-009-4 (ePub)
ISBN 978-1-78427-010-0 (Mobi)
ISBN 978-1-78427-029-2 (PDF)

British Library Cataloguing in Publication Data
A catalogue record for this book is available from
the British Library

Entypo pictograms by Daniel Bruce www.entypo.com

Cover image: Nataliya Hora/Shutterstock.com

Typeset by XL Publishing Services, Exmouth

CONTENTS

PART 2   USING YOUR DATASET – SUMMARIZING, VISUALIZING AND SHARING YOUR DATA

ABOUT THE AUTHOR

Mark Gardener (www.gardenersown.co.uk) is an ecologist, lecturer and writer working in the UK. As an undergraduate he studied a variety of natural science subjects including cell biology, physiology, biochemistry, chemistry, oceanography, geology, evolution and ecology. After graduation he carried out research in pollination ecology and has worked in the UK and around the world (principally Australia and the United States). Since his doctorate he has worked in many areas of ecology, often as a teacher and supervisor. Having dealt with so much data over the years he believes that it is time to help all scientists to arrange and manage their data more efficiently. This will make life easier for them, and their supervisors. It will also allow them to focus on what's ultimately important, conveying the information contained in their data to the world in general. He is currently self-employed and runs courses in ecology, data analysis, and R (a statistical programming language) for a variety of individuals and organizations. Mark lives in rural Devon with his wife Christine, a biochemist.

ACKNOWLEDGEMENTS

First of all I'd like to thank Nigel Massen at Pelagic Publishing for allowing me to sacrifice a few more trees in the advancement of knowledge. I'm also grateful to the Biological Recording programme, formerly at the University of Birmingham, now resident at Manchester Metropolitan University. Thanks especially to Sarah Whild for allowing me to be involved in a small way. The experience brought home to me how a solid foundation for recording data is important, not just in conservation biology but in all forms of scientific endeavour. Thanks also to Sue Townsend at the Field Studies Council for just generally being you. Finally, thanks to my wife Christine, who has to put up with me on a daily basis.

INTRODUCTION

Data are important if you are a scientist. You need data to answer research questions, test hypotheses and to undertake the process of science itself. Managing your data effectively is fundamentally important to making the scientific process run smoothly.

If you can manage your data effectively you can save time and effort, allowing you to focus on what is important: the patterns in your data. If you manage your data effectively you can explore your data more fully and produce high-quality graphs to help you and others understand it. Managing your data more effectively allows you to share your data more easily and communicate with others, a cornerstone of the scientific process.

Microsoft's Excel® spreadsheet is virtually ubiquitous and familiar to most computer users. It is a robust program that makes an excellent storage and manipulation system for many kinds of data, including scientific data. Excel is able to act like a database program and this ability gives it great power and flexibility. Using Excel you can explore your data and produce summary information necessary to help you in your research. You can also produce high-quality graphs, which can help you and others gain more insight into the patterns in your data. Using Excel you can save your data in a variety of formats, allowing you to share your data with other researchers and specialist analytical software as necessary.

WHO THIS BOOK IS FOR

The book is aimed at scientists at all levels. It is especially aimed at university-level students, from undergraduates to postdoctoral researchers. However, high school students should find the IT skills easy to acquire and the data management skills learnt will be a good foundation for any branch of further study. There is no assumption of prior knowledge although a passing familiarity with Excel would be useful.

Although aimed at scientists (the data examples are all scientific), any student who needs to use data should find this book helpful.

WHAT YOU WILL LEARN IN THIS BOOK

This book is intended to give you some insights into the processes involved in managing scientific data, allowing you to focus on your research questions. In particular you will learn how to:

- Maximize the usefulness of your data by maintaining a scientific recording format.
- Use your spreadsheet like a database.
- Manage your data, checking for errors and maintaining the most useful dataset possible.
- Summarize your data visually and numerically.
- Share your data and summaries (numerical and graphical).
- Ready your data for further analyses that Excel is unable to conduct.

In short, you'll learn everything you need to deal with data.

HOW THIS BOOK IS ARRANGED

This book is split into two parts, which cover issues surrounding the preparation and use of your data respectively. The first part is concerned with the preparation of your dataset and deals with the arrangement and management of your data. The second part deals with the use of your dataset and is concerned with ways to explore and share your data.

The first chapter deals mostly with showing you how to set out your data in a logical and useful manner. This systematic approach to data layout (let's call it scientific recording format) underpins everything else.

The second chapter deals with the building of your dataset and covers a range of data management issues, including:

- Adding to your data.
- Editing your data.
- Rearranging your data.

The third chapter deals with management issues relating to error checking, an important but often overlooked task.

The chapters in Part 2 deal with ways of using your data, such as:

- Visualizing your data – creating summary graphs and charts is important for you to make sense of your data and also to convey information about your data to others.
- Summarizing your data – lots of numbers are meaningless without interpretation and summary information. This summary information includes averages and correlations.
- Sharing your data – it is important to be able to share your data with others. You may also need to share your data with other computer programs so that you can carry out statistical analyses that Excel is unsuited for.

Throughout the book you will see example exercises that are intended for you to try out. In fact they are expressly aimed at helping you on a practical level – reading how to do something is fine but you need to do it for yourself to learn it properly. The Have a Go exercises are hard to miss.

Have a Go: Learn something by doing it
The Have a Go exercises are intended to give you practical experience at various tasks useful in preparing and managing your data. Many exercises refer to supplementary data, which you can get from the companion website.

↘ Go to the website for support material when you see the download icon.

Where you would be expected to type something, such as text or a formula, the text is shown in a fixed font like so: =AVERAGE(B2:B22). This should help you differentiate between names of variables, menu items and things you are expected to type yourself.

You will also see tips and notes, which will stand out from the main text. These are useful items of detail pertaining to the text which I feel are important to highlight.

Tips and Notes: Useful additional information
At certain points in the text you'll see tips and notes highlighted like this. These items contain things that are important to emphasize and mention especially.

At the end of each chapter there is a summary table to help give you an overview of the material in that chapter. There are also some self-assessment exercises for you to try out. The answers are in Appendix 1.

HOW TO READ THIS BOOK

You should start by reading Part 1 (chapters 1–3), since this is about preparing your dataset. Part 2 (chapters 4–9) is about using your dataset and you might want to dip into parts of this, according to the sort of data you have.

Chapter 4 shows how data can be in different forms, depending upon the nature of the study, so this is a good chapter to read through. Chapters 5–8 cover aspects particular to each different kind of data, so you might simply read the chapter(s) most appropriate for your data. Some of the processes you'll use will be common to more than one kind of dataset, so there is an element of repetition. The final chapter is about sharing your data so is relevant to all kinds of dataset.

SOFTWARE USED

Several versions of Microsoft's Excel spreadsheet were used in the preparation of this book. Most of the examples presented show version 2010 for Microsoft Windows®. You can see a summary of the differences between Excel 2010 and other versions (principally 2007 and 2013) in Appendix 2. There are also some notes about other spreadsheets.

SUPPORT FILES

The companion website contains support material including example spreadsheets containing data that are used for the examples and exercises. There are also some chart templates, which you can use to help you format certain types of graph (these are mentioned explicitly in the text).

↘ Go to the website for support material when you see the download icon.

The support material can be accessed here: www.gardenersown.co.uk/Education/ Lectures/index.htm

Most of the data examples were taken from the datasets accompanying R program: R Core Team (2012) *R: A Language and Environment for Statistical Computing*. R Foundation for Statistical Computing, Vienna, Austria. http://www.R-project.org/.

Other data came from some of my former students or were fabricated especially for this book.

Mark Gardener, Devon 2015

PART 1

Arranging and managing your data

WHAT YOU WILL LEARN IN THIS PART

Before you can undertake any scientific analyses you need to prepare your data. Your data need to be in an appropriate arrangement so that you can manage them more easily. Part of this data management includes editing, rearranging and checking for errors.

THE MAIN TOPICS ARE:

- How to arrange your data in scientific recording format.
- How to use your dataset like a database.
- How to add new variables to your dataset.
- How to make and use index variables.
- How to edit your data:
  - Using *Filter* tools.
  - Using *Sort* tools.
  - Using *Find and Replace* tools.
- How to rearrange your data.
- How to check your data for errors:
  - Using *Sort* and *Filter* tools.
  - Using pivot tables.
  - Using dot charts to help check your data for errors.
  - Using data validation tools.

1

ARRANGING YOUR DATA

All data are important. At the least they are important to you, as you've invested time and effort in collecting the data. Your data may well be important to others as well, it doesn't matter whether you are doing a high school project, a PhD or government research, the data you collect are important. You will use these data to help you make sense of your project and they may also be shared and presented to others. It is therefore important that your data are understandable by others. You may well take a break from your project so it is also helpful if your data are understandable by you when you return at some future date!

You may have spent a considerable time planning your work and deciding how to collect the data. You can also spend a lot of time collecting data so it is important to take care of them. The scientific process is a cyclical one and generally involves several stages:

- Planning.
- Data collection and recording.
- Analysis.
- Reporting and moving on.

You should have spent some time during the planning process to determine various aspects of your data:

- What data to collect.
- How to collect the data.
- How much data to collect.
- How to record the data.
- How to analyze and present the data.

In this book you will learn how to make best use of your data. The way you record your data underpins all your data management. It is easy to underestimate the importance of this aspect. Good data management can:

- Save time.
- Save money.
- Save effort.
- Reduce errors.

Good data management also means that you are able to add to your data at a later stage with minimal fuss.

You'll also learn how to explore your data and get some insights into the patterns that may (or not) exist in your carefully collected data. This aspect is sometimes called *data mining*, and can be a useful way to look for patterns and trends.

So, the way you arrange your data is of fundamental importance to your ability to utilize it. In the next section you'll see some examples of how you might set about arranging your data.

1.1 SYSTEMS FOR DATA LAYOUT

The way you arrange your data should be part of your general scientific approach. Science is a way of looking at the natural world. In short, the scientific process goes along the following lines:

1. You have an idea about something.
2. You come up with a hypothesis.
3. You work out a way of testing this hypothesis.
4. You collect appropriate data in order to apply a test.
5. You test the hypothesis and decide if the original idea is supported or rejected.
6. If the hypothesis is rejected, then the original idea is modified to take the new findings into account.
7. The process then repeats.

In this way, ideas are continually refined and our knowledge of the natural world is expanded. You can split the scientific process into four parts (more or less): planning, recording, analysing and reporting.

- *Planning*: This is the stage where you work out what you are going to do: formulate your ideas, undertake background research, decide on your hypothesis and determine a method of collecting the appropriate data and a means by which the hypothesis may be tested. This is the stage where you should be thinking about how to arrange your data to make maximum use of them.
- *Recording*: The means of data collection should be determined at the planning stage although you may undertake a small pilot study to see if it works. After the pilot stage you may return to the planning stage and refine the methodology. You collect and arrange your data in a manner that allows you to begin the analysis. The arrangement of your data should help you to check it for errors and also to add extra information at a later point. Good data layout also facilitates the following stages.
- *Analyzing*: The method of analysis should have been determined at the planning stage. You use analytical methods (involving statistics) to test a hypothesis. Having a good arrangement of data means that the analyses run smoothly.

- *Reporting*: Disseminating your work is vitally important. Your results need to be delivered in an appropriate manner so they can be understood by your peers (and perhaps by the public); this means summarizing your data numerically and graphically. Part of the reporting process is to determine what the future direction needs to be. Having a good data layout can help make your data understandable by others and also help you to present as usefully as possible.

Essentially you use the planning stage to help you determine what data to collect and how to arrange it. You use the recording stage to save your data in an arrangement that allows you to proceed to the analytical stage. The recording stage should also allow you to check for errors and permit you to add extra information that you may have overlooked at the earlier planning stage. If your data are arranged sensibly the analysis and reporting stages are facilitated, and you can share your data with others more easily.

It is easier to understand the issues by looking at some examples. In the following sections you'll see examples of different ways to set out data.

1.1.1 Common ways to lay out data

How you set out your data depends somewhat on the kind of analysis you are going to do. In the following examples you'll see several kinds of experimental situation.

Comparing samples

When you are comparing samples of things the simplest way to set out data is sample by sample. In Table 1.1 you can see such a layout, where there are two samples of data.

Table 1.1 Mandible lengths (mm) for golden jackals (*Canis aureus*) from specimens at the British Museum of Natural History.

Female	Male
110	120
111	107
107	110
108	116
110	114
105	111
107	113
106	117
111	114
111	112

Each column represents the data from a separate *sample*: there is one for females and one for males. The numbers in the columns show the lengths of the mandibles in millimetres. There is seemingly no great problem with Table 1.1 and if this were annotated fully it would certainly be acceptable as a data format. If you have more samples you simply have more columns, such as in Table 1.2.

Table 1.2 Fly wing lengths (mm x 10) for individuals fed on various sugar diets (C = control, G = glucose, F = fructose, F+G = fructose and glucose, S = sucrose).

C	G	F	F+G	S
75	57	58	58	62
67	58	61	59	66
70	60	56	58	65
75	59	58	61	63
65	62	57	57	64
71	60	56	56	62
67	60	61	58	65
67	57	60	57	65
76	59	57	57	62
68	61	58	59	67

In Table 1.2 there are five columns; one for each diet. The values in each column show the wing lengths of flies fed on that particular diet. If you had carried out this experiment and recorded the data in your lab notebook you would probably have included the date and a few additional details so that you could repeat the experiment at some future date if required.

As your data becomes more complicated it becomes more difficult to maintain a sensible layout (Table 1.3).

Table 1.3 Incidence of warp-breakage for two types of wool at different tensions.

A			B		
H	L	M	H	L	M
36	26	18	20	27	42
21	30	21	21	14	26
24	54	29	24	29	19
18	25	17	17	19	16
10	70	12	13	29	39
43	52	18	15	31	28
28	51	35	15	41	21
15	26	30	16	20	39
26	67	36	28	44	29

The data in Table 1.3 show the number of breaks in a fixed length of wool for three tensions (high, medium and low); there are two types of wool (A and B). In this instance the data are set out in two separate blocks, each corresponding to wool type. Whilst this makes some sense, you can see that if additional factors were to be included, things might get a little tricky to represent using the "sample" approach. In the following example the data are in three separate blocks because there are too many factors to show easily (Table 1.4).

Table 1.4 Abundance of beetle species at two kinds of habitat at three sites.

	Lower		Middle		Upper	
	Closed	*Open*	*Closed*	*Open*	*Closed*	*Open*
Sp.A	4	95	30	158	0	3
	9	71	45	120	10	9
	34	39	30	27	6	28
Sp.B	5	6	5	81	15	21
	3	0	0	51	2	25
	7	48	30	44	17	35
Sp.C	23	84	21	115	23	45
	12	75	44	110	1	43
	0	53	18	112	15	51

The data in Table 1.4 show the abundance of three species of beetle. The beetles were counted at three different sites. At each site there were two contrasting habitats. You can see that this layout has stacked blocks of data (the species); if there were any more variables you would not be able to display them sensibly in a single table.

Association analysis

In some kinds of analyses you might collect your data in the form of a *frequency table* (sometimes called a *contingency table*). An example is shown in Table 1.5 where you can see frequencies of combinations of hair and eye colour for female university students.

Table 1.5 Frequency of hair and eye colour for female students at the University of Delaware.

Hair/Eye	Blue	Brown	Green	Hazel
Black	9	36	2	5
Blond	64	4	8	5
Brown	34	66	14	29
Red	7	16	7	7

You can see from Table 1.5 that there are two sets of categories: the columns are for eye colour and the rows for hair colour. Each cell of the table shows the number of people (the frequency) with a particular combination of hair and eye colour.

The contingency table in Table 1.5 is certainly how you would set out your data in order to carry out the association analysis, but it might not be the most flexible approach, as you will see later. If you have more than two categories (perhaps male and female students) you would not be able to show the data in a single table.

Correlation and regression

When you are looking for relationships between factors your data tends to come in a particular way that lends itself to a certain layout. In simple *correlation* you have two variables and are looking to examine the strength of any link between them. In Table 1.6 you can see an example of the link between height and weight for a group of American women.

Table 1.6 Height (inches) and weight (lbs) of American women aged 30–39.

Height	Weight
58	115
59	117
60	120
61	123
62	126
63	129
64	132
65	135
66	139
67	142
68	146
69	150
70	154
71	159
72	164

The data in Table 1.6 are naturally arranged in pairs: a particular height is matched up with a particular weight (they are measurements from the same person of course). The arrangement is neat: each row represents data from one case (called an *observation* or *replicate*) and each column represents a *variable* (also called a *factor*). If the experiment becomes more complicated it is easy to add extra columns for the additional variables. In Table 1.7 you can see part of a larger dataset containing four environmental variables from New York.

The data in Table 1.7 are part of a larger set; there are 111 observations altogether. Each row shows the measurements for a day; each column shows measurements for a single environmental variable. The first column shows a simple index value here but it could easily have shown the actual date of the observations (see Section 1.4.2).

In most cases you have a *response variable* (also known as the *dependent variable*) and a number of *predictor variables* (also called *independent variables*). In Table 1.7 the *Ozone* variable is the response variable. The other variables are the predictors; you hope that they will help to predict ozone levels.

Table 1.7 Daily environmental measurements from New York, May to September 1973 (not all data are shown).

Obs	Ozone	Radiation	Temp	Wind
1	41	190	67	7.4
2	36	118	72	8.0
3	12	149	74	12.6
4	18	313	62	11.5
5	23	299	65	8.6
6	19	99	59	13.8
7	8	19	61	20.1
8	16	256	69	9.7
9	11	290	66	9.2
10	14	274	68	10.9

It is probably best to use the terms *response* and *predictor*, rather than *dependent* and *independent*. When you have multiple variables you do not really know if the predictor variables are actually independent of one another.

In many cases you are looking to see what effect the various predictor variables have on a particular response variable; in Table 1.8 you can see such a situation; an example of where you would use a *regression* analysis.

Table 1.8 Yarn breakage under load cycles (Cycles = number of cycles until failure, Length = specimen size in mm, Amplitude = time in min of loading cycle, Loading = load in g).

Cycles	Length	Amplitude	Loading
674	250	8	40
370	250	8	45
292	250	8	50
338	250	9	40
266	250	9	45
210	250	9	50
170	250	10	40
118	250	10	45
90	250	10	50
1414	300	8	40

The first column in Table 1.8 shows the response variable, the number of cycles of loading until the yarn breaks. The other columns show the predictor variables; they have an effect on the response variable.

The data examples you've seen here are all set out in subtly different ways. However, it is possible to use a single layout that permits you to represent all these experimental situations, as you will see next.

1.1.2 A standard layout for data

To be most useful your data has to be laid out such that it:

- Shows all the information.
- Is flexible.
- Can be checked for errors easily.
- Can be analysed easily.
- Can be extended and modified easily.

Meeting all those criteria could be a tall order but it is possible with a little thought. Look back at the jackal data in Table 1.1. There are two columns, one for each sample (male and female). This layout seems to meet the criteria. If you add more samples you can simply add more columns. Then the data would look more like the fly-wing data in Table 1.2, where there are five columns, one for each sample.

One potential problem with the multi-sample layout is that many computer programs cannot carry out an analysis with the data in this form. Another problem is shown by the next example shown in Table 1.3. Here you have the number of breaks in lengths of wool under three tensions. If you only had one sort of wool this would not be a problem but you have two sorts of wool. In order to display the data you have to make two blocks of results. In Table 1.3 the blocks are shown side by side but an alternative layout is where you have one above the other as shown in Table 1.9.

Table 1.9 Incidence of warp-breakage for two types of wool at different tensions. Alternative data layout with blocks stacked vertically.

Wool	H	L	M
A	36	26	18
	21	30	21
	24	54	29
	18	25	17
	10	70	12
	43	52	18
	28	51	35
	15	26	30
	26	67	36
B	20	27	42
	21	14	26
	24	29	19
	17	19	16

Continued

Wool	H	L	M
	13	29	39
	15	31	28
	15	41	21
	16	20	39
	28	44	29

Continued

This stacked-block layout has a certain logic: you can see the experimental situation fairly clearly. There are problems though; most computer programs cannot analyze the data in this form. The more additional variables you add to the situation the harder it becomes to display the data in a convenient manner. For example, Table 1.4, which shows the abundance of different beetle species, contains about as much information in a single table as you can get.

The key to arranging your data lies in being able to split it into two types: *variables* and *observations*. This is fairly apparent when you look at data like those in Table 1.8 for example. Each column represents a single variable. The first column is the response variable and the others are predictor variables, that is, the values in the predictor variables have an influence on the magnitude of the response variable. The rows of the dataset represent the individual observations (sometimes called replicates or *records*), and each row is an individual set of measurements. These come from your *sampling units*, which might be quadrats, subjects, time periods or whatever.

Sometimes the data do not fall neatly into response and predictor variables; the data in Table 1.7 are like this. None of the variables (columns) could be accurately described as response or predictor variables but nonetheless you are interested in their relationship to one another (you'd lean towards the *Ozone* variable being the response and the others predictors). The rows still represent the separate observations (daily throughout the experimental period).

The variables do not have to be numeric, so you can have columns like those shown in Table 1.10, which shows an alternative layout for the wool data shown in Tables 1.3 and 1.9.

In Table 1.10 the data are shown with a response variable in the first column; this is the number of breaks for lengths of wool. The next two columns show the type of wool and the tension. Both contain simple labels. Sometimes this kind of notation is called a *factor variable*, as opposed to a *continuous variable* (that is, a number). You should note however, that in some branches of science plain numbers are used as factors. This tends to occur in medical and psychological research, where it is felt (by some) that a plain number prevents any potential bias. Thus drug 1 might be a placebo whilst drug 2 has an active ingredient.

The rows of the dataset represent the individual observations, or replicates. Not all of the data are shown in Table 1.10 but you can look back to see that there are nine replicates for each combination of wool and tension (Table 1.9).

Table 1.10 Incidence of warp-breakage for two types of wool at different tensions. Alternative data layout with columns for variables and rows for observations (only part of the dataset is shown).

Breaks	Wool	Tension
26	A	L
30	A	L
54	A	L
25	A	L
70	A	L
52	A	L
51	A	L
26	A	L
67	A	L
18	A	M
21	A	M

Most computer programs are set up to expect data in this form and if you are going to carry out much in the way of serious analysis this is the way to arrange your data. Even if you are only going to use your spreadsheet for analysis this kind of layout is advantageous. You are essentially setting out your data like a database, a form that your spreadsheet can utilize.

In fact setting out your data in database form is the key to maximizing its flexibility and usefulness, as you will see shortly.

1.2 RECORDING FORMAT

It does not matter what kind of data you collect, but how you record your data is of fundamental importance to your ability to make sense of them at a later stage. If you are collecting new data you can work out the recording of the data as part of your initial planning. If you have past data you may have to spend some time rearranging them before you can do anything useful with them. The time you spend planning your data layout will pay handsomely later on.

It is easy to write down a string of numbers in a notebook. You might even be able to do a variety of analyses on the spot; however, if you simply record a string of numbers and nothing else you will soon forget what the numbers represent. Worse still, nobody else will have a clue what the numbers mean and your carefully collected data will become useless.

All recorded data need to conform to certain standards in order to be useful at a later stage. First of all you need to be able to make sense of them: if you cannot remember what a column of values represents you are in big trouble. It would be even better if anyone could understand what your data represent. In most cases you add data to a spreadsheet so that each column represents a variable. Somewhere you need to add notes so that a reader can see what the variables are, such as the units and how the data were collected (see Chapters 2 and 9). Each row should be a separate observation.

Some data may be date-sensitive so it is important to get into the habit of recording the date (see Sections 1.4.2 and 2.1.1). In some branches of science the location is especially important (for example, archaeology, conservation and ecology) so you may record a GPS location as well as the site name. In some disciplines the identification of an object or species is important so the name of the person who did the identification should be given.

> **Note: Biological records**
> In the conservation sector the recording of species information is especially important. *Biological records* have a layout that includes certain information; usually this can be thought of as the *who, what, where* and *when*. The *who* is the name of the person recording the data (and generally identifying the species), the *what* refers to the species recorded, the *where* includes a grid reference and the *when* is of course the date. Other information may be recorded, such as the number of individuals, the sex or life stage.

Your data are important. In fact they are the most important part of your research. It is therefore essential that you record and store your data in a format that can be used in the future (by you or others). There are some elements of your data that may not seem immediately important but which nevertheless are essential if future researchers need to make sense of them. Having complete scientific records is important because it:

- Allows the data to be used for multiple purposes.
- Ensures that the data you collect can be checked for accuracy.
- Means that you won't forget some important aspect of the data.
- Allows someone else to repeat the exercise exactly.

If you fail to collect complete data, or fail to retain and communicate all the details in full, your work may be rendered unrepeatable and therefore useless as a contribution to science.

So, have a column for every variable, including details of date and location. Have a separate row for each observation. Once your data are compiled in this format, you can sort them by the various columns, export the grid references to mapping programs, and convert the data into tables for further calculations using a spreadsheet. They can also be imported into databases and other computer programs for statistical analysis.

In fact Excel (and other spreadsheets) can use data in this scientific recording format like a database, enabling you to access your data in useful and meaningful ways.

1.3 TURNING YOUR DATA INTO A DATABASE

If you lay out your data in an appropriate layout you can use them in the most flexible manner. Excel (like other spreadsheets) has various tools that allow you to explore and visualize your data, helping you to carry out your analyses, check for errors and share your data and results with others.

The key to turning your spreadsheet data into a database is in using a layout in which each column represents a separate variable, whilst each row represents an individual record (that is, one row for each observation or replicate).

There are several Excel tools that you can use to help you manage your data; the most important is the PivotTable tool. A pivot table provides a way of managing and manipulating your data. You can use the PivotTable tool to arrange and rearrange your data in various ways that are useful, but hard to do with the spreadsheet directly. Most often the data are displayed in summary, the sum being the default (you can also show the mean). The more complicated your data are the more useful a pivot table becomes. You can use the PivotTable tool in many ways, such as to:

- Arrange your data in basic sample groups.
- Arrange your data in different groupings.
- View sample means or standard deviation.
- Visualize your samples using graphs.

The process of constructing a pivot table is fairly simple:

1. Click once anywhere in your block of data. There is no need to select any cells, Excel searches around the place you clicked and automatically selects all the occupied cells to form a block.
2. Click on the *Insert* tab in the Excel ribbon.
3. Click the *PivotTable* button on the *Insert* menu; this is usually on the far left.
4. A *Create PivotTable* dialog box opens. This allows you to select data and a location for the completed pivot table. The data are usually selected automatically (see step 1). Once you click *OK* you move to the next step.
5. From the *PivotTable Field List* task pane you build your pivot table. You drag fields, representing columns in your data, from the list at the top to one of the boxes at the bottom. These boxes represent the areas of the table.
6. Once the table is built you can use the *PivotTable Tools* menus (*Options* and *Design*) to customize your pivot table.

Once you have built a basic pivot table you can alter and customize it quite easily. You can drag fields to new locations and alter the general appearance of the table.

Being able to manage your data like this allows you great flexibility, especially with complicated data. In the following exercise you have a go at making a pivot table. You use the data on jackal mandible lengths, which you saw in Table 1.1. These data show the mandible lengths of male and female golden jackals. This example is about as simple as you can get so is a good starting point.

Note: Pivot tables in other spreadsheet programs
The examples in this book use Excel 2010 for Windows. Other versions of Excel also use pivot tables. In some versions of Excel the *PivotTable* button may be found in the *Data* menu. Other spreadsheet programs provide a similar tool (sometimes called DataPilot), often via the *Data* menu. Older versions of Excel (and Open Office™) use a slightly different way to construct pivot tables: you drag fields directly into the table. Although the details may differ the principle remains the same.

Have a Go: Make a pivot table
You will need the *jackal.xlsx* file for this exercise. The spreadsheet contains two worksheets: *Sample-layout* and *Recording-layout*. The first mimics the data in Table 1.1, whilst the second is in the more flexible *recording format*.

↘ Go to the website for support material.

1. Open the spreadsheet using Excel. You'll see the two worksheets, representing the data in two alternative layouts. Click the *Recording-layout* tab to go to the data in recording layout.
2. There are four columns. Column B contains the length, and column C contains the grouping variable (the sex, male or female). The other two columns are simple index values: column A is headed *Record*, whilst column D is headed *Observation*. These index columns are important as they help you group the data (see Section 1.4.1).
3. Use the mouse to click once anywhere in the block of data. It is not necessary to highlight any data; Excel recognizes data once you start making the pivot table.
4. Now click the *Insert* tab on the ribbon. Then click the *PivotTable* button. A dialog box entitled *Create PivotTable* opens (Figure 1.1) and you see that the data in the worksheet have been highlighted (you should be able to see the moving border, also known as marching ants).
5. You can choose where to place the final pivot table. Usually the button *New Worksheet* is selected, for most cases this is desirable. So, now click the *OK* button to create a new worksheet and start to build the pivot table.
6. You are now presented with a new worksheet and a task pane entitled *PivotTable Field List* (Figure 1.2). This task pane allows you to select the variables from your data; these are the *Fields*. You can see that there is a list of variables (fields) at the top. At the bottom are four areas; you create the pivot table by dragging the fields into these boxes.

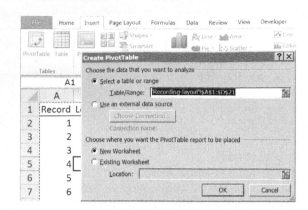

Figure 1.1 The *Create PivotTable* dialog box appears once you start the pivot table process.

Figure 1.2 The *Field List* task pane allows you to select variables and build a pivot table.

7. You must decide how to arrange your data; you can change your mind later on so there is no problem with some experimentation. For now use the mouse to click on the *Sex* item from the field list at the top, and then drag it to the box labelled *Column Labels*. You are going to set out the data in two columns, one for each sample (that is one for males and one for females).

8. The box labelled *Values* will contain the data so drag the *Length* field item from the list at the top to the *Values* box at the bottom. You will see that the pivot table incorporates the data immediately (Figure 1.3). At the moment the pivot table shows *Sum of Length*: you are seeing the total of each sample.

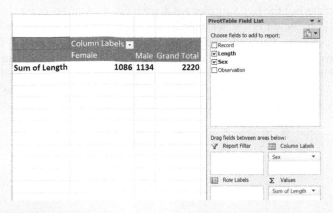

Figure 1.3 The pivot table incorporates data immediately as it is used so you can see the table forming.

9. Now the sum might be a useful summary statistic in some instances but for the time being you are more interested in reconstructing the sample data from the recording format. You need to use an index value to split the observations. Try dragging the *Record* field into the *Row Labels* box. The pivot table updates and you'll see the individual data (Figure 1.4).

	Sum of Length Column Labels			
	Row Labels	Female	Male	Grand Total
1	1		120	120
2	2		107	107
3	3		110	110
4	4		116	116
5	5		114	114
6	6		111	111
7	7		113	113
8	8		117	117
9	9		114	114
10	10		112	112
11	11	110		110
12	12	111		111
13	13	107		107
14	14	108		108

Figure 1.4 The choice of index field can affect the compactness of the layout. Here each observation has a unique index value.

10. This works but the two samples are displayed on separate rows (Figure 1.4). It would be better if the samples were side by side. The choice of index is important here. Click the tick-box beside the *Record* field item at the top of the *PivotTable Field List*. This will remove it. Alternatively you can drag the *Record* item out of the *Row Labels* box. Now select the *Observation* field item and drag this into the *Row Labels* box. The pivot table will update and now show you a more sensible layout (Figure 1.5).

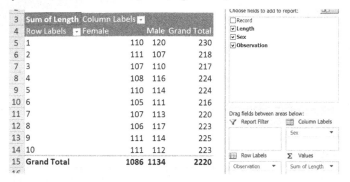

Figure 1.5 The choice of index field is important to get a compact table layout.

You have now completed a basic pivot table. There are many ways that you can tweak the table to alter the appearance of the layout, which you will discover later.

Once you have made a pivot table you can alter and rearrange it to suit your purposes. If you save the spreadsheet the pivot table is saved along with the rest of the spreadsheet.

If you click once in a pivot table the ribbon should display the *PivotTable Tools* menu (Figure 1.6). This has two parts; *Options* and *Design*; each contains tools that allow you to modify your pivot table. You can also right-click on many elements to bring up a context-sensitive menu.

Figure 1.6 Clicking on a pivot table activates the *PivotTable Tools* menu in the ribbon.

You can see from the preceding exercise that the way you index your data has an effect on the way you can display them in a pivot table. In the example the data are quite simple and only contain one predictor variable (*Sex*). As your data become more complicated it becomes more important to think about how to use index variables. Creating useful index fields is an important step in the planning process, allowing you to maximize the potential of your data.

1.4 PLANNING YOUR DATA LAYOUT

Planning how you are going to arrange your data is an important phase of scientific research. Whenever you collect data you should aim to save it in scientific recording format, with columns for the response and predictor variables, and rows for each observation. This layout permits the greatest flexibility, one of the benefits of which is that you can easily add variables at a later stage.

In order to make the best use of your data it is helpful to incorporate some index variables. These help you to use your data like a database and allow you to split, arrange and rearrange your data in meaningful ways. It is not usually essential to have these index variables to carry out the statistical analysis but they come into their own when you want to do some data mining, and start rummaging through your data looking for patterns and trends (see Part 2).

1.4.1 Index variables

The addition of index variables helps you to manage and manipulate the data, mostly using pivot tables. Excel has plenty of tools that can help you create useful index variables, as you'll see soon (Chapter 2). The more you learn about pivot tables the easier it becomes to plan useful index variables. Different kinds of dataset require different approaches to indexing:

- Association analyses involving contingency tables usually do not need any index variables.
- Correlations and regressions where the data are all continuous variables generally only require a simple row-index value. However, if there are grouping variables then index variables can be useful.
- When you have a factor variable (that is, the variable is in categories) an index can be based on the factor variable itself.
- When you have several factor variables they generally form a hierarchy; a basic index is usually based on the lowest level of the hierarchy (the individual observations). Additional (potentially more useful) indices can be built on higher levels of the sampling hierarchy.

Making an index variable involves counting the records that correspond to one of the predictor variables. Excel has a useful function, COUNTIF, which allows you to carry out this operation and so construct an index variable with minimal fuss. Essentially the COUNTIF function counts the number of items that match a criterion. You'll see more about the COUNTIF function and index variables later (especially in Section 2.1.1).

1.4.2 Using dates

The date is an important element in any scientific record, so you should plan to incorporate dates into your datasets. However, date formats can vary between computer programs so a useful approach is to record the date using several columns, one each for day, month and year (as numbers) and a fourth for the complete date (Table 1.11).

Table 1.11 Using separate columns for date elements can overcome issues with date formats between computers and programs.

Year	Month	Day	Date
2013	11	20	20/11/2013
2013	11	19	19/11/2013
2012	11	18	18/11/2012
2003	9	3	03/09/2003
1998	7	12	12/07/1998

The elements of the date (especially year and month) are generally useful as grouping variables. If you need a date column you can either enter the date separately or use an Excel function, DATE(year, month, day), where the year, month and day are the cells that hold the appropriate values.

If the date column appears strange and doesn't match the other three columns you know that something is wrong. If the date appears incorrect you will always have the original columns to use. You'll see more about date (and time) variables in Section 2.1.1.

1.5 USING RECORDING SHEETS AND NOTEBOOKS

Most of the time your data will be collected in the field or a laboratory and you won't enter them into Excel directly. You may get some of your data from equipment in the form of a printout or computer file. You may not even collect the data yourself; if you are running a large project you may have research assistants collecting data on your behalf (we can all dream).

It is important that you keep these sources of information organized. Loose pieces of paper have a habit of becoming disorganized very rapidly!

1.5.1 Data from equipment

If you are using a piece of equipment to collect data you'll get the information from it in one of three ways:

- From a display on a piece of equipment.
- From a printout.
- From a computer file.

Each of these scenarios requires a slightly different approach.

Data from equipment display

Many pieces of equipment do not give you a long-term result. Simply put, you have to record the result as you go. Such pieces of equipment include pH meters, rulers, weighing scales, thermometers and so on.

Inevitably this means that you'll have to write the result yourself in a notebook or onto a recording sheet.

Data from equipment printout

If the equipment you are using gives some kind of paper output you have several options. You can:

- Copy the result into a notebook or spreadsheet.
- Keep the printout as a record.
- Attach the printout to a notebook or recording sheet.
- Scan the printout and save the result in a computer file.

The approach you'll take depends somewhat on the size of the printout. If you have something small then copying the result into a notebook or spreadsheet is easiest. If the printout is somewhat larger you could stick it into your notebook (see Section 1.5.2). You might keep the printout separately, but this generally only works if the piece of paper is an appreciable size. Never try to keep loose paper! Use a ring binder to keep the printouts together. A more high-tech approach is to scan the printout and save the resulting PDF on your computer.

In any event you ought to transfer the results to your computer as soon as possible. You may want to have a separate series of folders for data pending; that is stuff that you've collected but not yet transferred to your computer.

Data from equipment computer file

If the piece of equipment saves results in an electronic format you can get your data directly. However, you may have to alter the format to get the data in an appropriate format for Excel.

Many types of equipment save their results in a format that Excel can read, such as CSV. You may have to do some rearranging and copy and paste will become a feature of your data entry process.

1.5.2 Data in notebooks

All scientists need a notebook. All the great scientists had great notebooks! You should use your notebook like you'd use a computer backup disk. If someone picked up your notebook they ought to be able to recreate your research.

What you record in your notebook will depend of course on the nature of your science. If you are an indoor scientist you should keep records of the solutions you made and the equipment set-ups you used for example. You'll need to keep notes about dates and times. Don't skimp on space; make fresh pages for new days. If you are an outdoor scientist you'll need to make notes about sites, equipment you used, experimental set-ups and so on. There really is little difference between indoor and outdoor scientists except that outdoor scientists tend to use pencils (you can write in the rain)!

You can use your notebook to store extra stuff, such as printouts from equipment. You can paste small items into place (the old-fashioned way!). You should be careful about this though, as such material has a habit of getting lost.

In many places the notebook is considered so important that it is not allowed to leave the laboratory. You should treat your notebook with equal care. You should try to transfer information and data to your computer as soon as you can. This gives you an opportunity to check your data as you enter it. You'll see more about error checking in Section 3.2.

1.5.3 Data in recording sheets

Recording sheets can be very useful, as you can make one as a template and distribute it to others (your research assistants). A recording sheet should contain spaces to record everything that you need, including the date and location. Recall the basic tenets of scientific recording: who, what, where, when. It is important to know which one of your research assistants collected the data because if there is a problem you'll be able to do something about it. It may be that the piece of equipment they were using has a systematic error and recorded data 10% lower than the real value. Not all mistakes are the fault of your research assistants!

Recording sheet design
The design of your recording sheet is very important. It is impossible to give anything other than vague guidelines because the design of your recording sheet depends upon the data you need to record. If possible you want to mimic the final spreadsheet. There will inevitably be variables that can be compressed to a single entry, such as the site name or date for example.

Make sure that you have a clearly labelled box for every required variable. Include the name of the recorder, even if it is yourself: if you lose a piece of paper it might find its way back to you if your name is on it. Having everything labelled ensures (as far as possible) that you record all the required information. It is especially important if someone else is carrying out the data recording. You may want to include some notes, which could be written on the back of the recording sheet, to help the recorder write the information as you want. For example, if you want the length of something, make sure that you specify the units and precision required. If you want centimetres with one decimal place (millimetre) precision then say so, for example, Length in cm (1 d.p).

In general colour coding is not overly helpful; the shading can simply make it harder to read an entry when you are transferring to Excel. If your recording sheet is set out sensibly it should be easy to find each element and enter the appropriate value into your main data spreadsheet.

Recording sheet storage
Recording sheets can be helpful to keep things neat and tidy. However, you'll have lots of separate pieces of paper to manage (note that "lots" is an anagram of "lost").

> **Tip: Outdoor recording sheets**
> Recording things outdoors leads to potential problems with the weather, and dirt. Using a pencil and a plastic bag overcomes some of the weather issues. Grubby fingers can leave marks on the recording sheet that can be misinterpreted. Decimal points can be a problem so avoid them altogether if you can; record in mm rather than cm with one decimal place for example. If decimals are unavoidable then ensure that all figures are written with them; for example, 20.0 rather than just 20. You can also use a slash instead for example, 20/0 as this is less open to (mis)interpretation on a page covered with mud splashes.

Don't try to keep recording sheets loose; they will simply end up getting out of order and you'll lose one along the way. Use a binder to keep them properly fixed together. You should to transfer the data from recording sheet to Excel as soon as you can.

Computers as recording sheets
In some cases you can use a computer directly in lieu of a recording sheet. Laptop and tablet computers are extremely portable and generally reliable. The advantage of using electronic recording is that you can transfer data to your main spreadsheet easily. The disadvantages are all those associated with any electronic device; battery or hard drive failure, water damage and being dropped on the floor!

If you do decide that a laptop or tablet computer is the way to go then keep things simple. Record your data into a spreadsheet that is separate from your main dataset. Use a spreadsheet like a recording sheet to minimize the typing required; there is no need to record the date more than once for example. You can then transfer the data later and look over it for consistency as you do so.

1.6   SAVING YOUR DATA

At some point you will need to save your data to a hard drive on a computer. Deciding how you are going to arrange your files is something you want to think about early, right at the planning stage! You'll need to make a folder to hold all the files used in your project. Keep the name short but meaningful. In this main folder you can keep together all the separate files (possibly in subfolders) that you are using in your project. These files might include:

- The main data spreadsheet as an XLS or XLSX file.
- Notes about protocols, methodology, site information and so on. These will probably be word processor files.
- Pictures, such as maps and photographs of equipment and sites. These will be in graphics formats such as JPG, TIF or PDF.

You will store your main dataset in your spreadsheet file but this file can contain more than just the data (you'll see how to add notes in Section 2.1.2). In general you want to keep any notes in your main spreadsheet as brief as you can.

In most cases you'll need additional files to keep detailed notes about methods, protocols, site photographs, maps and so on. You can keep a Notes worksheet in your spreadsheet and use hyperlinks in it to point to these separate files to make it easier to find what you need. Hyperlinks can be useful but use them sparingly; they should be there to help you and others navigate around the associated files.

Hyperlinks can be used anywhere in your spreadsheet. You can make a hyperlink from any cell in your workbook by using the *Insert > Hyperlink* button; this opens the *Insert Hyperlink* dialog box (Figure 1.7).

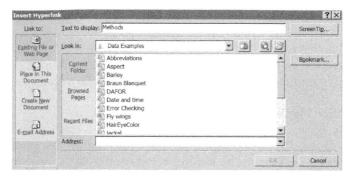

Figure 1.7 Hyperlinks can be used to point to files containing additional information.

The name of the hyperlink is taken from the contents of the cell; if you decide upon a different name you simply type it in the *Text to display* box and the cell contents are overwritten. This also means you can insert a hyperlink in an empty cell and add the text yourself. If you want to point to a specific point in the target file you need to select the file then use the *Bookmark* button. The target file will need to have bookmarks of some kind; if your target is a spreadsheet you can point to a specific worksheet or cell for example. You can also include a screen tip (a note that pops up when you hover over the hyperlink with the mouse) by clicking on the *Screen Tip* button.

Once you have your data saved on disk in a sensible structure you can manage your project more easily and effectively. An important element in saving your project data is the ease with which you can find the various pieces of information. You will need to find these for carrying out analyses and in writing up reports. You'll also need to be able to find information if you need to share your project and data with others. You'll see more about sharing your data in Chapter 9.

1.7   EXERCISES

1. Scientific recording format sets out data sample-by-sample, TRUE or FALSE?
2. Which of the following items is NOT a name given to individual measurements in your dataset?

    A) Record.
    B) Replicate.
    C) Item.
    D) Observation.

3. The COUNT function can be useful to help you make index variables, TRUE or FALSE?
4. Fill in the missing words: When you have your data in scientific recording format you can use your data like a ____, and the key to accessing the information is the Excel ____ tool.
5. The file *Fly wings.xls* contains data from an experiment to examine the effect of various sugar diets on fly growth. Open the file and look at the worksheet labelled *Scientific-format*. Make a pivot table and convert the data to sample format. What you are aiming for is shown in the *Sample-format* worksheet.

↘ Go to the website for exercise files.

The answers to these exercises can be found in Appendix 1.

1.8   SUMMARY

Topic	Key points
Data management	A good data management strategy can save you time and effort. It allows you to focus on the important things, such as the patterns in the data.
Sample format	In sample format you try to show each sample as a block. This can become unwieldy when you have more than one or two predictor variables. In addition, this format is not readily interpreted by most programs you'll use for statistical analysis.
Scientific recording format	In scientific recording format you set out your data so that each column is a separate variable. You have columns for the response variable and each of the predictor variables. Additional columns can be used for dates, names and index variables. Each row represents a single observation (also called a replicate or record). In this format (scientific recording format) your data can be used like a database, giving great power and flexibility.

Topic	Key points
Variables	Response variables are the variables that are affected by other things; they are sometimes called dependent variables. The variables that create the effect are called predictor variables (or independent variables). Continuous variables are numbers, which can take a range of values. Categorical or factor variables have fixed levels.
Index variables	Index variables allow you to differentiate between replicates from the same treatment (group of predictor variables). They are especially helpful when rearranging your data via a pivot table. Index variables can be used with any kind of data but are especially useful for differences data. The COUNTIF function can be used to help create index variables.
Types of dataset	You can think of data as being in one of three forms, according to the general kind of analysis you'll need to carry out: association data, correlation or regression data and differences data.
PivotTable	PivotTable is an Excel tool that allows you to arrange and rearrange your data in different ways in pivot tables. You can use a range of summary statistics (for example, average or standard deviation). When used with index variables you can present your data in sample format. There are four sections you can use to make a pivot table: *Row Labels*, *Column Labels*, *Values* and *Report Filter*.
Dates	It is helpful to split dates into their components (for example, year, month, day). The DATE function can "reassemble" these components into an Excel date.
Recording sheets and notebooks	Notebooks and recording sheets should support your main dataset. Try to transfer information from them into Excel as soon as you can. Don't keep loose sheets, bind them or paste into a notebook.
Saving data	Keep project files together, using folders and subfolders. The XLS or XLSX format for Excel files retains comments and text boxes as well as any pivot tables and charts. Additional notes and files can be saved to disk as separate files. You can insert hyperlinks to these items from Excel.

2

MANAGING YOUR DATA: BUILDING YOUR DATASET

You've seen in Chapter 1 that the most useful and flexible layout for your data is in *scientific recording format*, where each column represents a separate variable and each row is a record or observation. In this chapter you will start to explore the ways in which you can manage your data effectively. This includes adding new variables and information to your dataset (Section 2.1) and editing your data (Section 2.2). You'll also look at ways to rearrange data (Section 2.3), which can make things easier in the later stages of your project. Data management also includes checking for errors, which you'll read about in Chapter 3.

2.1 ADDING TO YOUR DATASET

When you add to your dataset you are adding either new rows or new columns. When you are adding rows you are generally adding new observations. When you are adding columns either you are entering recorded data or you are manufacturing a new variable. These manufactured variables can take various forms; you can think of them as being in three categories:

- *Index variables* – variables that you use to help you group and rearrange your data. You don't use index variables in any analysis although they can help to produce summary statistics and graphs.
- *Date variables* – variables with a special kind of formatting appropriate for dates. You can enter some date variables directly, like year or month. You can enter other date variables by combining simple date columns (such as those containing day, month and year information). You can use dates in later analysis.
- *Miscellaneous variables* – variables that do not fall into one of the previous categories, for example, a grouping variable that helps you to arrange your data, which might be used as a predictor variable in later analysis, or a label that helps you to group and visualize your data.

Excel has various tools that can help you create and manage these different variables.

2.1.1 Adding new variables

As part of the planning process you should have worked out what predictor variables will be included in your dataset. When you start the recording process, and create the initial spreadsheet, you will simply create new column labels so that you can keep track of these variables. As a general rule it is a good idea to put response variables in the first column(s), with predictor variables in subsequent columns. The last few columns can be reserved for additional information, such as index variables.

Index variables

Index variables allow you to sort and arrange your data in some meaningful fashion. Most usefully they work with a pivot table and allow you to extract your data from the main dataset and arrange them in sample groups. Some kinds of dataset do not need indexing: association data for example. Other kinds of dataset may need index variables to help arrange data into logical sampling groups:

- *Correlation and regression data* – sometimes benefit from index variables, usually if the dataset includes grouping variables.
- *Differences data* – always benefit from index variables, as the nature of the data is to split observations into sampling groups.

In general you should aim to have at least one index that reflects the lowest level of your sample hierarchy, the individual observations. This can be added to the data as new observations are entered into the spreadsheet. Other index variables can be created using the COUNTIF function in Excel. This makes it easier to manage, as you do not need to think what value needs to be entered for each observation, the formula takes care of it. Once you have entered some data the formula can be copied down the column and will automatically fill in the appropriate values.

The COUNTIF function has the structure:

```
COUNTIF(range_for_count, criterion)
```

The *range_for_count* is the range of cells that you want to count. The *criterion* is an expression that is evaluated; the value you get back is the number of cells in the range that match the criterion.

When you are making an index variable you usually want to match the contents of a cell, so you simply type the cell reference, for example,

```
COUNTIF(C$2:C12, C12)
```

In the preceding example your range of cells is C2 to C12: remember that the $ fixes a position, so if the formula is copied down a column the starting row is fixed (at 2 in this case). The criterion is the value of the cell at the end of the range. In other words, how many times you get a cell containing the contents of C12 in the block of cells from C2 to C12.

It is possible to use the criterion in other ways, for example if you wanted to see how many items in the range were greater than 50 you would type the following:

```
COUNTIF(C$2:C12, ">50")
```

Notice that the criterion is in quotes. You need to use quotes if the criterion is anything other than a cell reference or a plain number. Matching text is not case-sensitive so LOW is the same as low for example. You can also use the & symbol to join an element in quotes to a value; this is especially useful when matching cell contents using a cell reference.

In Table 2.1 you can see some examples of the use of the COUNTIF function. In all cases the cell range being investigated is C2:C12. Note that the $ has been used to fix the top end of the cell range, allowing you to copy the formula down the column.

Table 2.1 Use of criteria in the COUNTIF function.

Function	Meaning
COUNTIF(C$2:C12, C12)	Count the number of cells that have content equal to that of cell C12.
COUNTIF(C$2:C12, "Low") COUNTIF(C$2:C12, "LOW")	Count the number of cells that contain "Low" (not case sensitive).
COUNTIF(C$2:C12, 48) COUNTIF(C$2:C12, "48") COUNTIF(C$2:C12, "=48")	Count the number of cells that are equal to the value 48.
COUNTIF(C$2:C12, ">1")	Count the number of cells that contain a number greater than 1.
COUNTIF(C$2:C12, "<>1")	Count the number of cells with a value that is not equal to 1.
COUNTIF(C$2:C12, ">=5")	Count the number of cells that are greater than or equal to a value of 5.
COUNTIF(C$2:C12, "<=5")	Count the number of cells that are less than or equal to 5.
COUNTIF(C$2:C12, "<>"&C12)	Count the number of cells that are not equal to the contents of cell C12.

It is also possible to use wildcards to match any character(s). The query symbol ? matches any single character and an asterisk * matches several. So ???day matches words ending with "day" that have three letters at the beginning (for example, Monday, Friday, Sunday) but *day matches words that simply end with "day" (for example, all the days of the week). If you want to match a query symbol or asterisk then use the tilde, ~ as an escape character. So, ~* would match an asterisk, not any number of characters.

In terms of creating index variables you are most likely to use the simple cell reference (for example, first row of Table 2.1). Later on you'll see how to use the COUNTIF function to match criteria and make variables to be used for other purposes (miscellaneous variables).

Correlation and regression data

When you are dealing with correlations you usually have continuous numerical variables. Look back to Table 1.6 for example. There are two variables, *Height* and *Weight*. Each row represents the height and weight for an individual. In this case there is no pressing need for any kind of index variable. The data in Table 1.7 are more complicated because there are more columns of data but again there is no necessity for an index variable. The data includes a simple row index but it is not absolutely required. In both cases there are no grouping variables, so index variables are not useful.

Some data contain variables that represent categories (or can be treated as such); these are known as categorical variables. In such cases it would be useful to create an index based on a categorical variable. If there are several variables then you might make other indices. In the following exercise you can have a go at making index variables for some data that can be used for regression. The data are from the 1974 *Motor Trend* magazine and give fuel consumption and various other data for 32 makes of car. Your aim is to separate the data into samples.

Have a Go: Manage regression data
You'll need the *mtcars.xlsx* file for this exercise.

↘ Go to the website for support material.

Open the *mtcars.xlsx* file. It shows various models of car (from 1973–74) along with the fuel consumption (*MPG*, the response variable). The rest of the columns are data on various aspects of car design and performance. There are two worksheets; the first, *Raw-data*, is what you need to work on. The second worksheet *Indexed-data* contains index variables already worked out for you, so try not to cheat. In *Raw-data* you can see that some columns are akin to factors as they take simple integer values. These variables include the number of cylinders (*Cyl*), the transmission type (*AM*), the number of gears (*Gear*) and the number of carburettors (*Carb*). You could treat the number of gears as a continuous variable in a regression, but it is also obviously a potential category: you cannot get 4.5 gears for example. You could create index variables for any or all of these four variables but you'll focus on the number of carburettors to start with.

1. Navigate to the *Raw-data* worksheet. You are going to make an index based on the number of carburettors. To make things easier, start by sorting the data from lowest to highest carburettors. Note that this is not absolutely necessary but it helps you see what is happening in this case. Click once anywhere in the block of data then click the *Data > Sort* button on the ribbon.
2. From the *Sort* dialog box you need to select the sort criteria. Make sure the box that says *My data has headers* is ticked. Then select the *Carb* column from the *Sort by* box (Figure 2.1). Once you have done that you can click the *OK* button.
3. Now the data are rearranged in ascending order of number of carburettors. Click in cell L1 and type a name for your index: `Icarb` would seem suitable.
4. You could enter values manually but we'll use the useful COUNTIF function. In cell L2 enter the following formula: `=COUNTIF(K$2:K2, K2)`. Notice that you insert a $: this fixes the row.

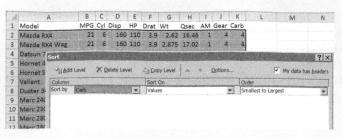

Figure 2.1 Rearranging data using the *Sort* dialog.

5. Now copy the cell L2 down the rest of the column. What the COUNTIF function does is to count how many cells in a range (the first part) match a criterion (the second part). As you move down the column the range of cells increases, so your index value increases. If you look at a cell lower down in the column you'll see the effect of using the $. The $2, marking the start of the range, has remained fixed but the other row numbers have altered; thus as you move down the column the range of cells is expanded.

6. Now you can make a pivot table and group the data into samples using the number of carburettors as the index values. Start the table by clicking once anywhere in the block of data, and then click the *Insert >PivotTable* button. Choose to place the pivot table in a new worksheet. The data should be selected automatically (check for the marching ants), so click on the *OK* button to make a new worksheet.

7. From the *PivotTable Field List* drag the variables to build the table. Start with the *MPG* field, which you should drag to the *Values* box. Now drag the *Carb* field to the *Column Labels* box.

8. You will see the table starting to form but at the moment it shows the sum of the mileage for each carburettor category. So drag the *Icarb* field to the *Row Labels* box. The table is now completed (Figure 2.2).

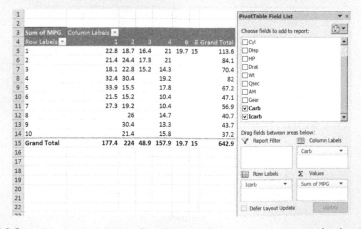

Figure 2.2 Completed pivot table made using an index variable to separate the data into samples.

Your pivot table is now completed, although there are some design elements that could be improved. You can now see that you have separated the data into several samples, each one representing the miles per gallon for a certain number of carburettors.

You could carry on and create index variables for other categorical variables.

Tip: Extending pivot tables
If you create a pivot table and later modify or extend the data you can update the information and alter the range of cells using the *Pivot Table Tools > Options* menu. The *Refresh* button updates calculations, whilst the *Change Data Source* button allows you to reselect the data to include any extra rows or columns.

Differences data

The kind of analyses where you are comparing groups of things lends itself to the use of index variables more than any other.

The simplest case involves comparing two samples. Recall the jackal data from Table 1.1, where there were two samples of mandible length (males and females). You looked at these data in recording layout where you ended up with two columns, one for *Length* (the response variable), and one for *Sex* (the predictor variable). There were ten measurements for each sex, so the index variable needed to reflect that (Table 2.2).

Table 2.2 Mandible lengths (mm) for golden jackals (*Canis aureus*) from specimens at the British Museum of Natural History. In recording layout.

Length	Sex	Observation	
120	Male	1	
107	Male	2	
110	Male	3	
116	Male	4	
114	Male	5	
111	Male	6	
113	Male	7	
117	Male	8	
114	Male	9	
112	Male	10	
110	Female	1	
111	Female	2	*Continued*

Continued	Length	Sex	Observation
	107	Female	3
	108	Female	4
	110	Female	5
	105	Female	6
	107	Female	7
	106	Female	8
	111	Female	9
	111	Female	10

So, for each level of the predictor variable you create unique index values; that is there are no repeated values within each group. When you have only two sample groups this is easy to do manually but you can still use the COUNTIF function. You can see the data and index in the file *jackal.xlsx*, which is on the companion website.

↘ Go to the website for support material.

One predictor variable
Working with many sample groups is nearly as simple as working with two samples. Look back at the data in Table 1.2: there are five samples. However, these are all measurements of the same thing (wing length) split according to the diet. So, they all represent a single predictor variable (the diet). If these data were laid out in recording format you would have a column for *Length* and a column for *Diet*. Your index variable would then correspond to the levels of the *Diet* variable. There are ten observations for each diet so your index would consist of the values 1 to 10 for each of those diets. You can see the data already indexed in the file *Fly wings.xls* (there are two worksheets, one in sample format, the other in scientific recording format), which is available on the companion website. The COUNTIF function was used to make the index variable in the same way as you saw earlier.

↘ Go to the website for support material.

Multiple predictor variables
When you have more than one predictor variable it can be harder to work out how to arrange your index variable. You may even require more than one index. Look back at Table 1.10, which shows data for the number of breakages in lengths of two types of wool under different tensions. The data in Table 1.10 are in recording layout: the first column (*Breaks*) is the response variable and the other columns (*Wool* and *Tension*) are predictor variables.

In most cases your predictor variables form a hierarchy. In the case of the wool example the individual measurements are at the bottom of the pile – they are the lowest in the hierarchy. The next level up is the *Wool* variable. Above this is the *Tension* variable and at the top is a combination of *Tension* and *Wool*.

You might be thinking that the hierarchy could just as easily be the *Tension*, followed by the *Wool*. It could be: in this case there is nothing to choose between them. There is perhaps something of a black art about working out a hierarchy. However, you can take much of the effort out of indexing by using the COUNTIF function. There are also other tricks that you can use to make life easier for yourself.

In the following exercise you can have a go at making alternative index variables to explore the hierarchy of predictor variables.

Have A Go: Explore a hierarchy of predictor variables and indices
You will need the file *warpbreaks.xlsx* for this exercise.

↘ Go to the website for support material.

1. Open the spreadsheet *warpbreaks.xlsx*. You want the worksheet entitled *Recording-layout*. You can see that there are three columns; the first is headed *Breaks* and is the response variable. The second column (B) is headed *Wool* and is a predictor variable with two levels, *A* and *B*, which represent the two different kinds of wool under test. The final column (C) headed *Tension* is also a predictor variable and gives the tension of the test run. The *Tension* variable has three levels, *H*, *M* and *L*. You are going to construct index variables. The second worksheet *Recording-layout-indexed* contains a completed version (try not to cheat).

2. Start by making a simple index to the unequivocally lowest level in the sampling hierarchy, the individual measurements. In cell D1 type a heading, Obs will do.

3. Enter a sequence of numbers in column D under your heading. You will need values from 1 to 54. Usually the easiest way to do this is to type the first two values, then highlight them and lastly position the mouse at the bottom right corner of the selection. The cursor changes from a fat cross to a thin cross and you can drag the selection down the column, filling in the sequence you started. If you cannot seem to get this to work, check the Excel options via *File > Options > Advanced*, where you can enable drag and drop (Figure 2.3).

4. Now you have one index variable. Make another one based on the next level of the hierarchy, *Wool* click in cell E1 and type a label, Wobs will do for this.

5. In cell E2 type a formula to start indexing: =COUNTIF(B$2:B2, B2). Don't forget that you need the $. You could use the mouse to click on the cells and then edit the formula afterwards (adding the $).

6. Now copy the cell in E2 down the rest of the column; this makes a complete index based on the *Wool* variable.

7. Click in cell F1 and type a label for the index based on the *Tension* variable: Tobs will do nicely.

8. Now click in cell F2 and enter a formula to start the index: =COUNTIF (C$2:C2, C2), then copy the formula down the column to complete the index for the *Tension* variable.

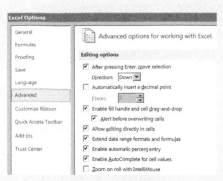

Figure 2.3 The Excel options allow cell drag and drop, useful for filling in sequences for index variables.

9. The highest level in the sampling hierarchy is a combination of the two predictor variables, *Wool* and *Tension*. You need to make an index for this but you will have to use an intermediate step. Make a label in cell G1, W:T will do nicely.

10. In cell G2 type a formula that combines the labels for the two variables, *Wool* and *Tension*: =B2&C2. The ampersand (&) simply joins two items together. Copy the formula in G2 down the rest of the column. You now have a label that shows the combination of the two predictor variables.

11. To make the index for the combination type a label in cell H1, Sobs will do. Then in cell H2 type a formula to start counting: =COUNTIF(G$2:G2, G2). Copy the H2 cell down the rest of the column to complete the index.

12. Now you have four index variables, *Obs*, *Wobs*, *Tobs* and *Sobs*. Click once in the block of data and start the pivot table process by clicking on *Insert >PivotTable*. Place the table in a new worksheet.

13. You can now explore the effects of using the different index values. Start by placing the fields as follows: *Tension* in *Column Labels*, *Breaks* in *Values*, *Wool* and *Tobs* in *Row Labels*. This makes a stacked table (Figure 2.4).

14. Try replacing the *Tobs* field with one of the other index variables. They all work adequately but the most efficient (fewest number of rows) are *Tobs* and *Sobs*, which are equivalent for this layout.

15. Now remove the index variable (the easiest way is to untick the box in the list of fields) and rearrange the table. Drag the *Wool* field out of the *Row Labels* section and into the *Column Labels* section so that it is above the *Tension* field.

16. Now the pivot table shows only the combined sum, so to restore the samples to the table you need an index variable. Try the various index fields in turn (they'll need to go in the *Row Labels* section), then end up with the *Sobs* field (Figure 2.5).

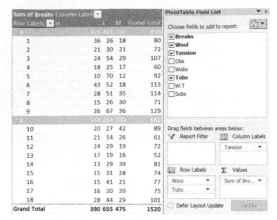

Figure 2.4 A stacked data table using one combination of variables.

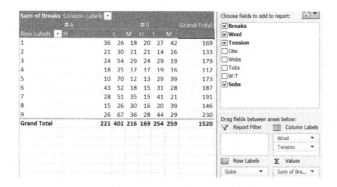

Figure 2.5 Predictor variables can be arranged in different locations to affect the layout of the final pivot table.

17. If you want to match Figure 2.5 more closely you will need to turn off the subtotals. Click once in the pivot table and click the *PivotTable Tools > Design > Subtotals* button, then *Do Not Show Subtotals* to remove all subtotals.

Try rearranging the fields and see the effects on the pivot table layout; there are many ways to arrange the data. You will have noticed that there is a second worksheet in the spreadsheet; this contains the index columns, already calculated for you.

Using index variables can be something of a daunting task; as you can see from the preceding exercise the more predictor variables you have the more indices you are likely to need. However, for most practical purposes you can simply use an index that relates to the rows of your data (the lowest level of the sampling hierarchy). This will allow you to recreate any of the samples but not necessarily in the most space-efficient manner.

> **Note: The & character**
> The & character is used for joining items, so = (A1&B1) joins the contents of cells A1 and B1. If you want to use a separator you simply include it inside double quotes, so = (A1&"-"&B1) joins the contents of cells A1 and B1 and places a dash between them.

Ideally you would set up your index variables at the planning stage. However, later in proceedings you may come up with new ideas about how to index your data; the recording format allows you to add new variables with minimal effort. Using the COUNTIF function makes it easy to set up index columns and you are able to extend the indices as you add more data simply by copying formulae down the appropriate columns.

Date variables

Most computer programs treat the date as a single numerical value from some starting point. For example, Excel for Windows uses January 1st 1900 as a starting point. If you enter the date as 01/01/1900 Excel stores the result as 1. Some computer programs use a different starting point. There is also potential confusion regarding the date format, so 05/07/12 might be regarded as 5th July 2012 in Britain but May 7th 2012 in the USA. It is a good idea to remove all doubt by creating multiple columns to hold the date, as you saw briefly earlier (Section 1.4.2).

> **Tip: Entering data in rows**
> When you have dataset out in recording format your variables are in columns and the individual observations (the records) are in rows. The default behaviour of Excel is to move to the next row down when you press Enter. If you use the Tab key you will move to the right (to the next column). When you press Enter you'll then go to the next row, in the column where you were when you first pressed the Tab key. This makes it really easy to enter data record by record.

Once you have separate columns for day, month and year, you can use a variety of Excel functions to deal with them. The most useful is the DATE function, which builds a date from the three separate columns:

```
DATE(year, month, day)
```

You give the cell references for the three date elements. When you enter this function the result is formatted in short date format, that is, the date elements are numbers separated by forward slashes. You can alter the date format by highlighting the cells and using the *Home > Format > Format Cells* button; a right-click also brings up a dialog box allowing you to *Format Cells*. From here you can alter the date format, choosing to display the month as text for example.

There are several functions connected to dates, as shown in Table 2.3:

Table 2.3 Some Excel functions for handling dates.

Function	Explanation
DATE(year, month, day)	Constructs a date object using three cells containing year, month and date respectively.
DATEVALUE (date_text)	Converts a text representation of the date to a number corresponding to Excel date value, where 1 Jan 1900 = 1.
YEAR(date_value)	Takes an Excel date value and returns the year as a number.
MONTH(date_value)	Takes an Excel date value and returns the month as a number.
DAY(date_value)	Takes an Excel date value and returns the day as a number.
TEXT(date_value, format)	Takes an Excel date value and returns a text result formatted in date-like manner.

These functions provide a good deal of control over dates, so it is worth having a little practice by doing the following exercise. A potentially useful variable is the *Julian day*: this is a single number that shows progress through a year. For every year January the first is represented as 1. Each subsequent day increases in value until you get to January the first for the next year, when the sequence begins at 1 again. In the following exercise you can have a go at calculating Julian days for various dates and also explore some of the other date functions.

Have a Go: Compute Julian day

You do not need a file for this exercise, as you'll create a new spreadsheet from scratch.

1. Open Excel and create a new workbook. In cells A1:C1 type headings for the date elements, Day, Month and Year.
2. In the first few rows type some values for dates, you can use any dates you like but make sure the first row is 1, 1, 1900.
3. In cell D1 type a label Date, because you are going to construct the date from the three date elements. In cell D2 type a formula to make the date: =DATE(C2, B2, A2). You can copy the formula down the column to display the date for all the rows you entered in step 2. Note that the displayed dates are in *date* format. Try altering the format to *general*, and see what happens, using the *Number* section in the *Home* menu.
4. In cell E1 type a label Text, because you are going to show the date in text format. In cell E2 type a formula to display the date in text format from the date you just computed: =TEXT(D2, "dd/mm/yyyy"). Now copy the formula down the column. Notice that the cells are not in date format but *general*.
5. In cell F1 and type a label XL Date, because you are going to display the date as a plain number (using January the first 1900 as a baseline). In cell F2 type

a formula to display the date as an index number: =DATEVALUE(TEXT(D2, "d/m/yyyy")). Copy the formula down the column. Notice that the cells are formatted as *general*. Try altering the format to *date* and see what happens.

6. In G1 and type a label &, because you are going to use the ampersand to make a text-like representation of the date. Now go to cell G2 and type a formula to join together the three date elements as plain text: =A2&"/"&B2&"/"&C2. You can copy the formula down the rest of the column. The cells appear like dates but alter the format and see what happens (nothing).

7. In cell H1 and type a label Build, because you are going to build a date index. In cell H2 type a formula to build a date index from the three date elements: =DATEVALUE(TEXT(A2&"/"&B2&"/"&C2, "dd/mm/yyyy")). Copy the formula down the column when you are done. Note that the cells appear as *general* format. Try altering the format to *date* and see what happens.

8. In cell I1 type a label Julian, because you are now going to work out the Julian day. In cell I2 type a formula to determine Julian day from the date in column D and the year in column C: =DATEVALUE(TEXT(D2,"d/m/yyyy")) - DATEVALUE(TEXT("1/1/"&C2, "d/m/yyyy")) + 1. It is quite long but you've seen all these elements already. The first part works out the index value for the entered date. You then work out the index value for January the first of that year and subtract it. You then add one at the end (because Jan 1st is 1). You can now copy the formula down the rest of the column. The cells are formatted as *general*.

9. Type labels in cells J1:K1, Month and Year, because you are going to extract the month and year from the index value in the *XL Date* column. In cell J2 type a formula to determine the month: =MONTH(F2). Now copy the formula down the rest of the column. Notice the format is *general*; altering it does not enable you to get the month as text.

10. Go to cell K2 and type a formula to extract the year from the date index: =YEAR(F2). Copy the formula down the rest of the column. Note that the results are in *general* format.

↘ A completed version of this exercise is available as part of the data on the companion website; the file is called *Date and time.xlsx*.

You have manipulated the date in several ways but have not actually entered the full date directly in any single cell. For some practice you might like to enter a column of dates and explore the various functions on them.

You've seen in the preceding exercise that there are several ways you can manage dates; the key is to keep the day, month and year separate because this gives you the greatest flexibility.

Note: Date format

The TEXT function allows you to fix a date format. The format is given in quotes for example, `"dd/mm/yyyy"`. If you use `"d/m/yy"` your days and months appear without leading zeroes (so January displays as 1 rather than 01) and the year will be in short format (for example, 2012 will appear as 12). Beware of short years because 01 might mean 1901 or 2001.

Time variables

Times are dealt with in a similar manner to dates. Excel handles times using a 24-hour system, and stores them as decimal fractions in the range 0–1 (to 5 decimal places). A time of 00:00:00 (midnight) is zero, whilst a time of 23:59:59 is just under 1. You create a time from separate columns using the TIME function:

```
TIME(hours, minutes, seconds)
```

You specify the cells containing values for the hours, minutes and seconds. The result is a cell with a *time* format. You can alter the format to display the result as a 12-hour format (that is, to include AM or PM) but the original hour cell has to be in 24 hour format.

Tip: Losing seconds

You may want to keep the time without having to enter seconds, that is, just using hours and minutes. If that is so then use the TIME function but use a zero in place of the seconds cell reference. The result displays the seconds as 00 (you can alter the format).

You can reformat a cell containing the time by right-clicking and selecting *Format Cells*, or use *Home > Format > Format Cells*. At first glance there are not many options for the time but if you select the *Custom* section you have a wider choice of format (Figure 2.6).

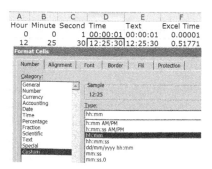

Figure 2.6 The *Custom* format section of the *Format Cells* dialog box gives a wide choice of formatting options for time (and others).

You can use the TEXT function in a similar manner to that which you used for dates, the difference being that you give a format relevant to times, for example, "hh:mm:ss". Some Excel functions for handling times are shown in Table 2.4.

Table 2.4 Some Excel functions for handling times.

Function	Explanation
TIME(hour, minute, second)	Constructs a time object using three cells containing hour, minute and second respectively.
TIMEVALUE(time_text)	Converts a text representation of the time to a number between zero and one (midnight is 0 and 23:59:59 is just under 1).
HOUR(time_value)	Takes an Excel time value and returns the hour as a number.
MINUTE(time_value)	Takes an Excel time value and returns the minute as a number.
SECOND(time_value)	Takes an Excel time value and returns the second as a number (from 0 to 59).
TEXT(time_value, format)	Takes an Excel time value and returns a text result formatted in time-like manner.

Excel uses 24-hour times so it is usual to record your data in the same manner, with hours in the range 0–24. You may want to record the time in 12-hour format and have an AM:PM indicator (as a separate column), or indeed you may be sent data in that format. It does take a bit of juggling but it is possible to convert 12-hour time to 24-hour. In the following exercise you can have a go at this for yourself.

Have a Go: Convert 12-hour times to 24-hour times
You do not need any data for this exercise, as you'll create a new spreadsheet from scratch.

1. Open Excel and start a new workbook. In cells A1:D1 type some headings; Hr, Min, A:P and Time will do nicely.
2. Now starting in row 2, enter some values for hours between 0 and 24 in the Hr column and some values for minutes between 0 and 59 in the Min column. In the A:P column type either AM or PM.
3. Go to cell D2 and type a formula to calculate the time: =IF(C2="PM", IF(A2<12, TIME(A2+12, B2, 0), TIME(A2, B2, 0)), IF(A2=12, TIME(A2-12, B2, 0), TIME(A2, B2, 0))). Now copy the formula down the column.
4. The times are displayed as a decimal so highlight the cells and alter the format to *Time*.

Essentially the formula adds 12 hours to times that have a PM indicator, as long as the times are not >12. Have a look at what happens when you create ambiguous times, like 12:00 AM and 12:00 PM or 00:00 AM and 00:00 PM. Try impossible times like 27:30 and see what you get.

↘ There is a completed version of this worksheet in the file *Date and time.xlsx*, which is available as part of the data on the companion website. The worksheet is named *12 to 24*.

You can see from the preceding exercise that it is a bit tricky to take into account the various possibilities. The formula needs to take into account the label (either *AM* or *PM*), so an IF function is used. However, it is possible that an hour value could be mistakenly entered as a 24-hour value. This means that you need another IF function to take care of this. A compromise is built into the formula in the exercise: an hour of 12 is taken as either midnight or noon and the label (*AM* or *PM*) determines how it is recorded as one or the other. A value of zero hours is more likely to be midnight but in this formula a label of *PM* will treat it as noon. You could nest another IF function into the formula but it is getting complicated enough! It might be better to leave it as it is and look carefully for 0 hour values when you do your error checking (see Chapter 3).

In the main it is easier to stick to 24-hour time formats if you can!

Miscellaneous variables

Normally you will have identified the main predictor variables at the planning stage, before you collect any data. If you carried out a *pilot study*, you might have thought of extra variables, but you can think of the pilot study as part of your planning process.

At the planning stage you may also have worked out some additional variables that you'd like to record, such as index values, dates and so on. Other ideas may occur to you as you are entering your data. These miscellaneous variables can be thought of as falling into several categories:

- *Binary* – that is, a variable that can only have two forms (often 0 or 1 but could be anything, male or female for example).
- *Categorical* – a variable with several (text) values.
- *Label* – a variable that you will use as a label; this is similar to a categorical variable.
- *Replacement* – here you replace the original data with something slightly different.
- *Abbreviation* – an abbreviated form of the original variable; this is similar to a replacement variable.

Some variables may fall into more than one category. In the following sections you'll see some examples of how to use spreadsheet tools to make creating and managing the variables in your data easier.

Binary variables

A binary variable is one that has only two forms. Often these forms are numeric, (0 or 1), but there is no reason why they cannot be anything. The binary variable could be a response variable, the presence or absence of something for example. These kinds of response variables tend to lead to analysis by logistic regression. The binary variable could also be a predictor variable and simply be entered into the dataset as you record the data.

On the other hand you may decide later that you would like to split your data into chunks according to some criteria, for example "Red" and "Other", so that you focus on one group.

The most practical way is to make these binary variables involve the COUNTIF function. In the following exercise you can have a go at making binary variables in two subtly different ways.

Have a Go: Make binary variables

You'll need the *mtcars.xlsx* file for this exercise.

↘ Go to the website for support material.

The file shows various models of car (from 1973–74) along with the fuel consumption (*MPG*, the response variable). The rest of the columns are data on various aspects of car design and performance. There are two worksheets; the first *Raw-data* is what you need to work on. The second worksheet *Indexed-data* contains an extended version with additional index variables.

1. Open the *mtcars.xlsx* spreadsheet and navigate to the *Raw-data* worksheet. Notice that column I contains a variable that is already binary, the transmission type (*Automatic* or *Manual*). Right now you want to look at the *HP* variable (column E); notice it contains numeric values.
2. You are going to "rate" the vehicles as powerful or feeble using a cut-off value from the *HP* column. So, click on cell L1 and type a label, Power for the new variable.
3. In cell L2 type a formula that inserts a 1 for cars with more than 100 horsepower and 0 for those with less: =COUNTIF(E2, ">100"). You can now copy the formula down the rest of the column.
4. Zeroes and ones are fine but sometimes you want a variable that is a bit more meaningful. This time you will make a binary variable that splits out all the Mercedes. So, click in cell M1 and type a label, Merc.
5. In cell M2 type a formula that examines the car name, returning Merc if that is found or Other if not: =IF(COUNTIF(A2,"Merc*"), "Merc", "Other"). Copy the formula down the rest of the column.
6. Make a pivot table from the new data and explore the two binary variables you just made. Try putting the *Power* field into the *Column Labels* area, the *Cyl* field in the *Row Labels* area and the *HP* field in the *Values* area. The table shows the sum of *HP*, which is not very helpful. Click on the *Sum of HP* field in the *Values* area to bring up a menu (Figure 2.7), and click on *Value Field Settings*.

Figure 2.7 Clicking on a field brings up a context-sensitive menu.

7. When the *Value Field Settings* dialog box opens select *Average* from the *Summarize value field by* list and click *OK* (Figure 2.8).

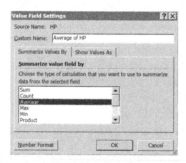

Figure 2.8 The *Value Field Settings* dialog box allows alternative summary statistics.

8. Being able to summarize by a mean value is very useful (the AVERAGE function works out the mean). Tidy up the pivot table by removing the subtotals; use the *PivotTable Tools > Design* menu and the *Subtotals* button.

9. Return to the *Value Field Settings* dialogue box by clicking on the *Average of HP* field and selecting *Value Field Settings*. Use the *Number Format* button and change the format to two decimal places. Now you can see how the power of the vehicles changes with numbers of cylinders (Figure 2.9).

Average of HP	0	1
4		76.33 111.00
6		122.29
8		209.21

Figure 2.9 Completed pivot table using a binary variable and an average (mean) summary.

What happens if you set the *HP* of a car to exactly 100? The binary variable you made in steps 2–3 records 0. You can shift a 100-HP car into the upper category by modifying the formula. Change >100 in the formula to >=100 to do this.

The IF function in step 5 allowed you to provide two options: what to do if the condition was true and what to do if the condition was false.

Note that the asterisk was used as a wildcard in step 5. Effectively this matched the beginning of the car name.

Altering the summary statistic in a pivot table to a mean value (the AVERAGE function) is generally a lot more useful than the sum. You'll see rather more of this kind of thing when you come to exploring and summarizing your data (see Part 2).

Categorical variables

A categorical variable is one that has two or more distinct levels, as opposed to a continuous variable where you have a number. In the previous section you saw a special case of the categorical variable, a binary variable that had only two levels.

You can use categorical variables (sometimes called *ordinal* variables when they can be ranked) for various purposes; most commonly you'll use them as predictor variables and enter them directly when you record your results. You can also use a categorical variable as a label; you'll see this in the following section. The label may or may not be used as part of the analysis.

At other times it can be useful to take a continuous variable and split it into chunks. An example of this is aspect, where you take a compass reading in degrees and convert to an ordinal aspect (NW, N, W, SE and so on). In the following exercise you can have a go at making a categorical (ordinal) variable using aspect.

Have a Go: Make a categorical variable from a continuous variable
You do not need any data for this exercise, as you will make a spreadsheet from scratch.

↘ There is a completed version on the website; the file is called *Aspect.xlsx*.

1. Open Excel and start a new workbook. In cell A1 type a label, Aspect, this will be the main variable. In the cells underneath type a few compass values to represent some aspect data, so you need numbers from 0–359 (360 is 0, that is, North).
2. Now in cell B1 type a label CQ, to hold the compass quadrant result (the categorical variable).
3. In cell B2 type a formula to convert the continuous compass bearing to a quadrant using 8 cardinal points of the compass: =FLOOR(((A2/22.5) + 1) / 2, 1).
4. Now copy the formula down the rest of the column.

You have now converted the continuous variable *Aspect* to an ordinal (categorical) variable *CQ*. The FLOOR function rounds numbers down. The 22.5 part splits the compass data into eighths. The 1 part at the very end uses one significant figure.

There is a completed version of the spreadsheet, *Aspect.xlsx*, on the companion website. This contains several worksheets (you will do more work with aspect later) and the compass quadrant calculations are in the *Quadrant* worksheet.

↘ Go to the website for support material.

It is possible to devise all manner of ways to produce categorical variables. Here you converted a numerical variable into another numerical variable, albeit an ordinal one. Later you'll see how to make ordinal variables containing text values.

Label variables

Sometimes it is useful to make label variables to help group your data or as an intermediate step in another calculation. You saw this earlier when you used the data on wool breakage to combine two predictor variables.

The ampersand character & is used to join items in Excel functions. You can specify various elements:

- Cell ranges – given as A3, B9 and so on.
- Numbers – given as plain numbers, 0, 18, 4.6 and so on.
- Text – given in quotes, "-", "/" and so on.

In order to make simple labels you don't need any functions but you need to type the = to tell Excel that you want to create something. Then you specify the elements you want, separated by & symbols.

In the example file *warpbreaks.xlsx* the original levels of the two variables *Wool* and *Tension* are very short, so the resulting combination is also short. If you have long names and need to shorten your labels you can use some of the functions that Excel uses to handle text items. You'll see more about making abbreviations later but right now you'll focus on the LEFT and RIGHT functions.

The LEFT function takes a *string* (a series of text characters) and extracts the leftmost part of it as a new item. You specify how many characters you want to extract:

```
LEFT(text_string, characters)
```

The RIGHT function operates in exactly the same way but returns the number of characters you specify from the rightmost end of the string:

```
RIGHT(text_string, characters)
```

You can use LEFT and RIGHT in conjunction with & to construct label variables from longer items. In the following exercise you can have a go for yourself.

Have a Go: Use text functions to make label variables

You'll need the data file *Barley.xlsx* for this exercise. The file contains data from yields from several varieties of barley grown at different sites in 1931 and 1932. You want the worksheet *Barley*. The other worksheet, *Barley-Labelled* contains a version with the labels (and an index column) already calculated – try not to cheat.

↘ Go to the website for support material.

1. Open the *Barley.xlsx* spreadsheet. You can see that there are four columns; *Yield, Variety, Year* and *Site. Yield* is the response variable and the rest are predictors. Go to cell E1 and type a heading `Si-Yr` for the label variable you are going to make.
2. Click in cell E2 and type a formula to create a label from the *Site* name and *Year*. You want the first letter of *Site* and the last two characters of *Year*: `=LEFT(D2, 1) & "-" & RIGHT(C2, 2)`.
3. Now copy the formula in cell E2 down the rest of the column.
4. Type a heading in cell F1 `Va-Si` for the label you are going to make from *Variety* and *Site*.
5. In cell F2 type a formula to make a label from *Variety* and *Site*. You'll need to use both ends of *Variety* to get unique names: `=LEFT(B2, 2) & "." & RIGHT(B2, 2)`.
6. Now copy the formula down the rest of the column.

The label variables can be constructed in various ways; the other worksheet, *Barley-Labelled* shows the three predictor variables combined in all four possible ways.

You will often use these made-up variables to help you group and organize your data. The shorter labels can be especially helpful when it comes to visualizing your data, to help prevent cluttering graph axes (see Part 2).

Replacement variables

Sometimes it is awkward to get the LEFT and RIGHT functions to work out appropriate labels. You may want to alter a variable to make a label that bears little resemblance to the original. In such cases you can create *replacement variables*. Essentially you take the original value and swap it for something else. One example might be Roman numerals, where you wish to convert a regular (Arabic) number into a Roman numeral.

Another example are abundance scales; these are common in ecology. Rather than measure the abundance of an organism exactly you record the scale equivalent. Such scales are sometimes called *ACFOR* or *DAFOR* scales; the letters represent an abundance class. So, A stands for abundant, D for dominant and so on. The *Domin scale* is another example where the percentage cover of plants is converted to an ordinal scale in the range 0–10.

If you want to convert a regular number to a Roman numeral you can simply use the ROMAN function, which produces a text value:

```
ROMAN(value, form)
```

The *value* part is usually a cell reference but you can type in a number directly or include a formula that will produce a number. The *form* is a number in the range 0–4, which gives increasingly concise versions of the Roman numerals; this can be helpful for large numbers. You can also specify TRUE to get the classic Roman numeral or FALSE to get the most concise version.

If you have Excel 2013 or Open Office (or Libre Office) version 4 then you have access to the function ARABIC, which allows you to return a number from a Roman numeral:

```
ARABIC(text)
```

You simply give the Roman numeral and the result is an Arabic number. Usually you give a cell reference but you can also specify the text directly (in quotes). The function is not case sensitive, so lowercase letters are fine.

If you do not have the later version of spreadsheet, then you need a different approach. What you can do is to make a table that contains original values and their replacements. This is known as a *Lookup Table*.

There are two main functions that use lookup tables; VLOOKUP and HLOOKUP, which work in similar ways:

```
VLOOKUP(match, table, replacement, approx)

HLOOKUP(match, table, replacement, approx)
```

The VLOOKUP function reads down the first column of data in the lookup table until it matches what you asked for, and then it takes a replacement value from another column. The HLOOKUP function reads along the first row of your lookup table until it matches what you requested, then it takes a replacement from another row in the lookup table. You can also specify if you want the match to be exact or approximate: you set the *approx* part to TRUE for approximate (this is the default if you do not specify) or FALSE otherwise.

So, you specify what you want to match, this is usually a cell containing a value you wish to replace. You then specify the location of the lookup table and the column, or row, of the lookup table that you want to act as the replacement value. In general your lookup table should be sorted in ascending order (using the first column). This allows the approximate matching part of the function to operate correctly. If your lookup table is not sorted you can set the *approx* part to FALSE. This has big implications, as you'll see in the next two exercises.

In the following exercise you can have a go at using the VLOOKUP function.

Have a Go: Use a lookup table to make replacement variables

You do not need any data for this exercise as you will create a spreadsheet from scratch.

1. Open Excel and start a new workbook. Go to cell A1 and type a heading Obs, to top a simple index of observations. Now in cell B1 type a heading Abund, to top a column of abundance values, recorded using a DAFOR scale.
2. Type a series of numbers in the *Obs* column, 1–10 will be sufficient. Now go to column B and type some values to represent abundance values in a DAFOR scale (D = dominant, A = abundant, F = frequent, O = occasional, R = rare). Start with D, A, F and so on. Then use lower case letters so that you end up with ten values.
3. In cell C1 type a heading Num for the column that will contain a numeric equivalent for the abundance scale.
4. Before you complete column C, you'll need to make a lookup table to hold the values of the original abundance scale and the corresponding replacement values. So, go to cell E1 and type a heading Scale. In the cells underneath type the values of the abundance scale; E2 will contain D, E3 will contain A and so on.
5. In cell F1 type a heading Ordinal, to remind you that this column will hold an ordinal value, relating to the abundance scale. Enter the numbers 5 to 1 in cells F2:F6. You should now have a lookup table with one column containing the original DAFOR labels and one containing corresponding numerical values.
6. Now return to cell C2 and type a formula that looks to match an entered abundance in the lookup table and returns a numerical replacement: =VLOOKUP(B2, $E$2:$F$6, 2). Note that you need the $. If you use the mouse to select the cells you can edit the formula afterwards.
7. Copy the formula in cell C2 down the rest of the column. Notice that the lower case letters are matched but that the abundance "A" produces an error #N/A for both upper and lower case (Figure 2.10).

	C2			f_x	=VLOOKUP(B2,$E$2:$F$6,2)		
	A	B	C	D	E	F	G	H
1	Obs	Abund	Num		Scale	Ordinal		
2	1	D	5		D	5		
3	2	A	#N/A		A	4		
4	3	F	3		F	3		
5	4	O	2		O	2		
6	5	R	1		R	1		
7	6	d	5					
8	7	a	#N/A					
9	8	f	3					
10	9	o	2					
11	10	r	1					

Figure 2.10 A lookup table must be sorted into alphabetic order if the approximate matching is TRUE (the default).

8. The problem is that the VLOOKUP function is trying to find an approximate match. Click on cell C2 and edit the formula: add an extra parameter, FALSE at the end so the function reads: =VLOOKUP(B2, $E$2:$F$6, 2, FALSE). Copy the function down the rest of the column. Your values should now be represented correctly (Figure 2.11).

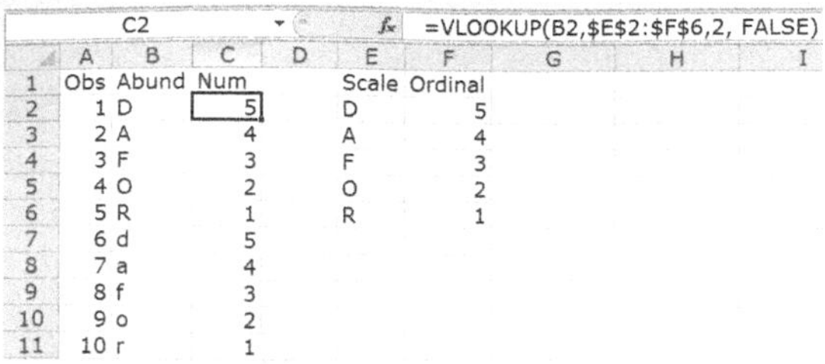

	A	B	C	D	E	F	G	H	I
1	Obs	Abund	Num		Scale	Ordinal			
2	1	D	5		D	5			
3	2	A	4		A	4			
4	3	F	3		F	3			
5	4	O	2		O	2			
6	5	R	1		R	1			
7	6	d	5						
8	7	a	4						
9	8	f	3						
10	9	o	2						
11	10	r	1						

C2 fx =VLOOKUP(B2,$E$2:$F$6,2, FALSE)

Figure 2.11 Using FALSE at the end of a VLOOKUP function allows the lookup table to be unsorted.

9. Now highlight the values in the lookup table (cells E2:F6), then use the *Home > Sort & Filter > Sort A to Z* button. The lookup table should now be sorted alphabetically.

10. Click in cell C2 and edit the formula to either remove the FALSE and the preceding comma) or alter FALSE to TRUE. Now copy the formula down the rest of the column. Highlight the lookup table cells (E2:F6) and sort them from Z to A. Now you get errors again.

11. The lookup table does not have to be in the same worksheet. Highlight the lookup table (and the headings) in cells E1:F6 and use the *Cut* button, this is on the *Home* menu and looks like a pair of scissors. Now make a new worksheet using the icon at the bottom (or use the *Home > Insert > Insert Worksheet* button). Then paste the cells into the new worksheet using *Home > Paste.*

12. You can rename the worksheet containing the lookup table; right-click the appropriate Tab and select *Rename*. If you now return to your original data and look at cell C2 you'll see the reference to the cells of the lookup table appear with the name of the worksheet and a following exclamation mark (Figure 2.12)!

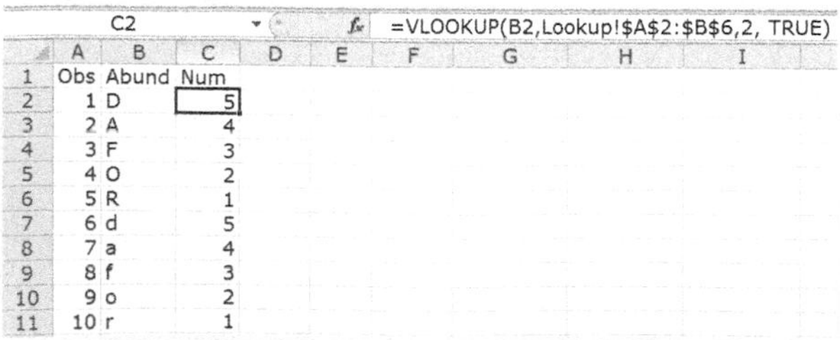

	A	B	C	D	E	F	G	H	I
1	Obs	Abund	Num						
2	1	D	5						
3	2	A	4						
4	3	F	3						
5	4	O	2						
6	5	R	1						
7	6	d	5						
8	7	a	4						
9	8	f	3						
10	9	o	2						
11	10	r	1						

C2 fx =VLOOKUP(B2,Lookup!$A$2:$B$6,2, TRUE)

Figure 2.12 A lookup table can be in a separate worksheet.

Having a lookup table in a separate worksheet is often preferable to having the lookup next to the data. You may want more than one lookup table and it is handy to have them together in one place.

↘ The companion website contains a completed version of the spreadsheet from this exercise, called *DAFOR.xlsx*.

In the preceding exercise you converted a simple DAFOR scale to a numerical equivalent. The opposite would involve converting a numerical scale into an ordinal equivalent. This is where having approximate matching becomes particularly useful.

The Braun-Blanquet scale is used to convert percentage vegetation cover to an alternative (ordinal) representation. Table 2.5 shows one version, which you'll use in the exercise that follows.

Table 2.5 A version of the Braun-Blanquet scale.

Cover in %	Scale
> 75	5
51–75	4
26–50	3
6–25	2
1–5	1
< 1	+

In Table 2.5 any cover greater than 75% is recorded as a five. A cover of greater than half but less than three quarters is recorded as four, and so on. Anything less than 1% is shown as a plus sign (zero cover is either not recorded or shown explicitly as 0).

In the following exercise you can have a go at using approximate matching to convert a value recorded as a percentage into a Braun-Blanquet scale equivalent.

Have a Go: Use approximate matching in a lookup table
You do not need any data for this exercise as you will create a spreadsheet from scratch.

1. Open Excel and start a new workbook. Go to cell A1 and type a heading Percent, to indicate that this column will contain values in percentages (i.e. 0–100). Type in a few values in the cells underneath your heading, so that you have some data to work with.

2. In cell B1 type a heading Braun-B, to indicate that this column will contain values using the Braun-Blanquet scale.
3. Go to cell D1 and type a heading Orig, then in E1 type Repl. These are the headings for the lookup table, which you will make next.
4. Now click in cell D2 and type the value 0 (zero). In cell E2 type the value 0 (zero) again. Now you have the first row of the lookup table, a value of zero will be replaced with zero. Easy so far.
5. Go to cell D3 and type the value 0.001. In cell E3 type a +. You now have the second row of your lookup table. A very small value (0.001) will be replaced by a plus symbol.
6. Now go to cell D4 and type the value 1 (one). In cell E4 type the value 1 (one). In cell D5 type the value 5 (five), and in cell E5 type the value 1 (one). Rows 4–5 give the lower and upper limits for the Braun-Blanquet scale value 1 (that is, 1% to 5%).
7. Go to cell D6 and type the value 6. In cell E6 type the value 2. In cell D7 type 25 and in cell E7 type 2. You have now set the boundaries for the Braun-Blanquet scale value 2.
8. Carry on in the same manner for the other Braun-Blanquet scale values. Your final lookup table should look like Table 2.6.

Table 2.6 Excel lookup table for conversion of percentage values to Braun-Blanquet scale.

Orig	Repl
0	0
0.001	+
1	1
5	1
6	2
25	2
26	3
50	3
51	4
74	4
75	5
100	5

9. Once the lookup table is completed you can start using it to convert percentage data into Braun-Blanquet scale equivalents. Go to cell B2 and type a formula to look up a value in the *Percent* column and replace it: =VLOOKUP(A2, $D$2:$E$13, 2). Don't forget to use the $ symbols to

fix the boundaries of the lookup table. You might prefer to use the mouse to select the cells and edit the formula afterwards, adding $ as needed.

10. Now copy the formula in cell B2 down the rest of the column.

Try typing different percentage values into the *Percent* column. See what happens especially when you are close to a boundary. Try values of 49, 50, 50.5 and 51 for example. You'll see that the 50.5 is still matched to 50, so returns a Braun-Blanquet value of 3 but once you get to 51 you switch to the next scale value. This is because the VLOOKUP function is using approximate matching and 50 is a closer approximation to 50.5 than 51.

↘ The companion website contains a version of the spreadsheet used in this exercise; it is called *Braun Blanquet.xlsx*.

Note: Approximate matching in lookup tables
Using approximate matching can be very useful. However, it can lead to unexpected results if you are using text values, so you need to proceed with caution.

Since you should be recording your data in scientific recording format you are unlikely to need the HLOOKUP function. You should have all your variables in separate columns, so the VLOOKUP function should be sufficient. The HLOOKUP function works in exactly the same manner except that you now read across rows to match items and select replacements.

Abbreviations

There are many occasions when it would be useful to use an abbreviated name in your data. Short versions of names are especially useful in graphs, helping to reduce clutter. Biological species names in particular tend to be quite long. It is possible to type the abbreviated name in directly as you enter data of course but this means entering the name twice in effect.

Abbreviations using lookup tables

One way to proceed is to use replacement variables. You can easily make a lookup table that contains the original names and their abbreviations. You can even enter your data as an abbreviation and use a lookup table to insert the full name. This can be helpful if you have many species or species with long and complicated names, as it reduces typing errors (see Chapter 3).

You can use a pivot table to help you make your lookup table. Use the names as the rows of the pivot table, which you can then copy to the clipboard. If you add more data later you can re-create your pivot table and check for additional names. The pivot table should be sorted in alphabetical order, but you can always check your lookup table and apply a sort to be on the safe side.

You should proceed along the following lines:

1. Make sure your data are in the correct layout: you'll need to have a single column containing the *names* you want to replace.
2. Start by making a pivot table. Click once anywhere in the block of data and then use the *Insert > Pivot Table* button.
3. Choose to place the pivot table in a new worksheet and then click *OK*.
4. Now drag the *names* field from the list at the top of the *PivotTable Field List* to the *Row Labels* box at the bottom. Your pivot table now begins to form and you'll see a list of *names* in alphabetical order.
5. Use the mouse to highlight the *names* then copy to the clipboard. Click the icon at the bottom to make a new worksheet and then paste the names starting at cell A2 (to leave room for headings).
6. You can now type abbreviations into column B as appropriate.
7. Once you have the lookup table you can return to the main data and use the VLOOKUP function in a new column.

When you use the Paste operation you are copying not only the contents of the cells but also the formatting. In some cases, especially with pivot tables, this can include unwanted formatting. This hidden information can sometimes hinder you by doing unexpected things. For this reason it is useful to use additional Paste options, often referred to as *Paste Special*. The *Paste* button can be expanded (there is a triangle symbol, Figure 2.13) to show additional options. The options you see depend on the version of Excel and if there is any data in the clipboard.

You can also click *Paste Special* to bring up a more exhaustive dialog box (Figure 2.14).

Figure 2.13 The *Paste* button can be expanded to show additional options.

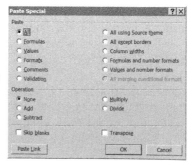

Figure 2.14 The *Paste Special* dialog box gives a range of options for the paste operation.

A right-click will also show the *Paste Special* option; if there is data in the clipboard you'll also see the common formatting options in Excel.

Typing abbreviations manually can be somewhat tedious, so in the next section you'll see how to use Excel text functions to automate the process.

Abbreviations using text functions

You saw previously how to use LEFT and RIGHT functions to get portions of text for label variables. You can also use these functions with others, to help you construct abbreviations. There are five main text functions; these are shown in Table 2.7.

Table 2.7 Some Excel text functions that can be used to help make text abbreviations.

Function	Explanation
LEFT(text, num)	Returns a number of characters from the leftmost part of the text you specify.
RIGHT(text, num)	Returns a number of characters from the rightmost part of the text you specify.
MID(text, start, num)	Returns a number of characters from the text you specify, starting from a point you select (as a numerical starting point).
FIND(what, where, start)	Returns the position of a text string (what), within another (where). You can also specify the starting point (the default is 1).
LEN(text)	Returns the number of characters in the text you specify (usually a cell reference).

You use the & character (ampersand) to join together the various portions of the abbreviation. The functions themselves are fairly easy to understand but it can take a bit of thought to combine them in an appropriate manner. You can have a go at making abbreviations in the exercise that follows shortly.

An abbreviation needs to have several characteristics to be most useful:

- *Short* – there is little point having a long abbreviation!
- *Unique* – each abbreviation needs to refer to a single original; there is no point having two names with the same abbreviation.
- *Meaningful* – ideally you should be able to recognize the original name from the abbreviated form.

Scientific names for species are often long and it is therefore often desirable to have a shortened version. Scientific names are generally given as a pair of names, with a *genus* and a *trivial* name. This is often called the *scientific binomial naming* system. If you take the first three letters of the genus and the first three letters of the trivial name you are likely to have a unique abbreviation in most cases. If this 3+3 system does not give unique names adding a character to one or both names will usually be sufficient.

It is easy enough to get the first three letters of the genus: as this is at the start of the name the LEFT function will work for you. You now need the first three letters of the trivial name but the start of this could be at any point in the name. The trick is to search for the space between the genus and trivial names. This gives you the location for the start of the trivial name (actually one place to the left of the correct start point). You can then use the LEFT function to take the three letters of the trivial name, which you can stitch onto the first part using the & symbol.

In the following exercise you can have a go at making some abbreviations.

Have a Go: Use text functions to make abbreviated names

You'll need the file *Abbreviations.xlsx* for this exercise. It contains a few species names in scientific binomial format. Or you can simply type some into column A of a new worksheet (put a label Species in cell A1). The file contains two other worksheets, *Simple-Abbr* and *Complex-Abbr*, which contain completed formulae.

↘ Go to the website for support material.

1. Open the *Abbreviations.xlsx* file and navigate to the *Names* worksheet. This contains a few species names, as scientific binomial, in column A.
2. In cell B1 type a heading for the abbreviated names, Abbr will do.
3. Go to cell B2 and type a formula to take the first three letters of the genus and the first three of the trivial name, using a full stop as a connector:
   =LEFT(A2, 3) & "." & MID(A2, FIND(" ", A2) + 1, 3).
4. Once you are happy with the formula in step 3 you can copy it down the rest of the column.

The formula has three sections. The first takes the leftmost part of the name. The middle part simply tacks on a full stop. The last part is more complicated: it uses MID to extract text from the middle of the whole binomial name. The FIND part looks for the position of the space and uses that value to extract the three letters of the trivial name.

The completed version of the worksheet is entitled *Simple-Abbr*. The *Complex-Abbr* worksheet has a more complicated example, which you'll see shortly.

If you need a different abbreviation structure to that demonstrated in the preceding exercise, you can simply edit the formula.

> **Tip: Simple abbreviation formula**
> If you make a generic formula you can store it in a text file and paste it as required. For example:
>
> =LEFT(cr, 3)& "." & MID(cr, FIND(" ",cr) + 1, 3)
>
> where cr is the cell reference.
> You can edit the formula to suit.

If you have more complicated names you can extend the idea. For example, in ecology it is common to record some plant species with the name of the vegetation layer they were found in, for example.

- *Quercus robur* (s)
- *Salix caprea* (c)
- *Hedera helix* (g)
- *Urtica dioica*

The first is an oak that is found in the shrub layer. The second is a willow found as a canopy tree. The third is ivy, found in the ground layer. The last is stinging nettle and no layer information is given.

The basic approach will not work here; you may have plants recorded from a non-specific layer and any plant recorded from more than one layer would be duplicated. You need a more complicated formula with two options:

```
IF(RIGHT(cr, 1) =")", LEFT(cr, 3) & MID(cr, FIND(" ", cr) +
    1, 3) & "." & MID(cr, FIND("(", cr) + 1, 1), LEFT(cr, 3) &
    "." & MID(cr, FIND(" ", cr) + 1, 3))
```

The first part checks to see if there is a closing parenthesis. This would indicate that the layer information was present. If that is TRUE then the next part is carried out. This makes the 3 + 3 abbreviation as before but does not separate it with a full stop. Instead the "." is added after the trivial portion and the character after the "(" is inserted.

The second part of the formula is carried out if there is no ending ")" (that is, there is no layer information). This is shown in **bold** to make it easier to see. The formula would produce the following results if applied to the species shown earlier:

- Querob.s
- Salcap.c
- Hedhel.g
- Urt.dio

The *Abbreviations.xlsx* spreadsheet, found on the companion website, contains the worksheet *Complex-Abbr*, which shows this formula in action. You can copy and paste this to make things easier!

↘   Go to the website for support material.

Tip: Checking for unique abbreviations
Check that your abbreviations are unique from time to time. The simplest way is to make a pivot table. Put the original names and the abbreviated names in the *Row Labels* part of the pivot table. Then use *PivotTable Tools > Design > Report Layout > Show in Compact Form*. You'll need to switch off subtotals using the *Subtotals* button in the *PivotTable Tools > Design* menu. This allows you to check for duplicates more easily.

Using a formula can save you a lot of typing but with a large dataset you might consider it worthwhile to take a composite approach.

1. Use a pivot table to make a list of species names.
2. Copy the species list to a new worksheet.
3. Use a formula to make abbreviations for your list, thus making a lookup table.
4. Use the lookup table from step 3 to create the abbreviated names in your main data. Set approximate matching to *FALSE*.
5. If you find non-unique abbreviations you can alter the lookup table.

If you add a new row to your data and the species abbreviation returns as #N/A it means that the species was not in the lookup table (or you spelt it incorrectly). If you need to add a new species to your lookup table do **not** add a row at the end. Insert a row in the middle of the table. This will update any formulae that refer to the lookup table, so you will not need to alter them. Once you've added the appropriate names to the lookup table you can sort it alphabetically. Now the new species abbreviation should appear correctly.

2.1.2   Adding notes to your data

It is important that anyone can look at your data and understand what it represents. You may take a break from your project or be working on more than one thing, so it is important that you can understand what your data represents! Whilst there is some personal satisfaction in carrying out an investigation, you'll eventually need to tell someone else what you've done. This can be very informal, such as a chat with colleagues. On the other hand it may be more formal and involve a report or presentation.

It is important that alongside the actual data you have notes about how things were carried out. You should make a note about the equipment you used and solutions that you made and used. These details are important because they allow someone else to be able to repeat your work. This repeatability and transparency is an important aspect of scientific research.

You can keep many notes right alongside your data in the spreadsheet. Of course the spreadsheet is not a word processor but it is useful to keep basic notes tied to your data, rather than in a separate file.

Notes about variables
You can make simple notes about the individual variables by using the *comment* feature. If you place a text *comment* (called a *note* in some spreadsheets) in the cell containing the column heading you can include any relevant information. You place a comment by right-clicking on a cell and choosing *Insert Comment* from the popup menu. You can also insert and manage comments using the *Review* ribbon (Figure 2.15).

Figure 2.15 The *Review* ribbon allows insertion and management of comments, as well as other features.

A cell that contains a comment is indicated by a small red triangle in the top right corner of the cell. If you hover over the cell with the mouse, the comment appears. If you right-click the cell you'll be able to edit or delete the comment.

You can alter the way Excel displays comments from the *File > Options > Advanced* dialog box. You need to scroll down to find the *Display* section; the default option is usually *Indicators only, and comments on hover*.

In general you want to keep the comments short but meaningful (Figure 2.16).

Figure 2.16 A comment pops up when the mouse hovers over the cell containing the comment.

If you save your data as a plain text CSV (see Chapter 9) the comments will be lost, so make sure that you have the information elsewhere.

Notes about protocols and methods
The basic cell comment is most useful when it is kept short, for notes about units for example. If you need to enter notes about methods and protocols it would be better to use a different method.

Notes in cells
You could simply type notes into the cells of a worksheet. However, they do not always display very neatly. There are three main options:

- Type the note into the cell as it is.
- Type the note then set the wrapping so the text appears over several lines (that is, the cell height increases).
- Type the note then highlight several cells to merge and wrap.

These options affect how the cell appears (Figure 2.17).

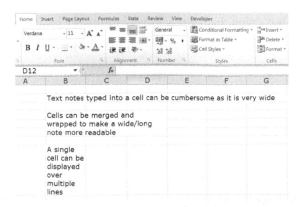

Figure 2.17 Text typed into a cell can be wrapped and merged to alter its appearance.

You can alter the wrapping and merging of cells using the buttons in the *Alignment* section of the *Home* menu. These options are also available if you right-click and select *Format Cells* or use the *Home > Format > Format Cells* button. You can control the width of cells by dragging the right boundary of the column heading, right-clicking the column header (the part with the letters, not row 1), or by using *Home > Format*. You can also set the column width to match the contents by double-clicking on the right boundary of the column header. You can select several columns and a double-click (at the right boundary in the header) will set all columns to optimal width.

In general trying to add lots of text to cells as comments or notes is somewhat unwieldy.

Notes in text boxes

Excel provides a range of drawing tools; these are available via the *Insert* menu on the ribbon. In older versions of Excel or other spreadsheet programs you can find the drawing tools in a separate toolbar (try *View > Toolbars*). The basic tools in Excel include *Shapes* and *SmartArt*. The *SmartArt* tools include graphic objects that allow you to visualize hierarchies and processes, which can be helpful. These *SmartArt* objects are undoubtedly very clever but there are two problems:

- *SmartArt* takes time and effort to learn.
- *SmartArt* is not compatible with other spreadsheets.

If you are not bothered about compatibility then have a look at the *SmartArt* options.

A more robust alternative is to use simple shapes and text boxes. Simple text boxes are generally adequate to keep your notes. To make some text-box notes use the following process:

1. Use the *Home > Insert > Insert Sheet* button to make a new worksheet to hold your notes.
2. Right-click the name of the new sheet, in the tab at the bottom of the screen. Then *Rename* the worksheet to something meaningful. Notes seems appropriate.
3. If necessary click and drag the tab to a new location in the list of sheets at the bottom.
4. Insert a text box using the *Insert > Shapes* button; the basic text box is in the *Basic Shapes* section (Figure 2.18).

Figure 2.18 Text boxes are found in the *Insert > Shapes* dialog box, in the *Basic Shapes* section.

5. The cursor changes to a dagger-like symbol; click where you want the top-left of the box to start.
6. Start typing. The text box expands as you type. Alternatively you can click and drag the box to the size you want.

You can set the box width using the mouse. The width will remain fixed at what you set but if you add more text the box will expand downwards to accommodate it. If you want to make the box a different width, use the mouse again.

If you click a text box the *Drawing Tools* menu appears, which allows a range of formatting options (Figure 2.19).

Figure 2.19 Clicking on a drawing object (for example, a text box) activates the *Drawing Tools* menu.

You can alter the appearance of your text box in many ways but using *Shape Fill* and *Shape Outline* are the most basic. You can alter the text colour and font by selecting it in the box and using the *Font* section of the *Home* menu (try right-click also). It is probably best not to apply too much formatting; the notes are there to help you and others understand the data.

Notes about observations

Sometimes you want to add a note about a specific observation in the dataset. Maybe you were unhappy about the reading or there was something notable you want to record.

You can use a comment in the cell of the first row of that particular observation; this is especially useful if your first column is a simple index variable. On the other hand you can use a column expressly for comments. It is most useful if this is the final column, as the cell could be quite wide.

Notes to accompany your data

You won't want to put everything in the same file as your dataset. Some of your notes and observations will be too long to fit sensibly. Keep your notes in the main file short and to the point. Other items can go in separate files.

It sounds obvious but you should make a folder on your computer to hold all the files from a single project. You can have subfolders for different topics, like recipes for solutions, site information, species photographs and so on. Of course you'll also have a folder for the data.

Keeping to a system allows you to keep things organized. Arranging your files and information is an important part of the planning process. You should regard this as integral to your project and not just a boring administrative process. If you are organized right at the start you can focus on your data and the administration is actually a lot less tedious!

2.2 EDITING YOUR DATA

Once you have entered some data in your spreadsheet you may want to undertake some minor editing. You'll look at error checking your data later (Chapter 3) but here you'll get a flavour of some of the spreadsheet tools that can help you carry out some minor editing tasks.

The three main tasks you'll see described here are:

- *Find and replace* – this allows you to make minor changes to your data.
- *Sorting data* – this can help you rearrange your data into sensible order, making it easier to make minor edits and to spot errors.
- *Filtering data* – filters allow you to view chunks of your data, making it easier to make minor edits and spot errors.

These editing tasks overlap somewhat with the things you'll do when error checking (Chapter 3) and also in rearranging data (Section 2.3), which you'll see later.

2.2.1 Find and replace

You can search the contents of cells and optionally replace them using the *Home > Find & Select* button. This brings up various options, *Find* is what you generally want since you can access *Replace* from *Find* and you'll need to find something before you can replace it!

When you click *Find* the *Find and Replace* dialog box appears. Click on the *Options >>* button to see the various options (Figure 2.20).

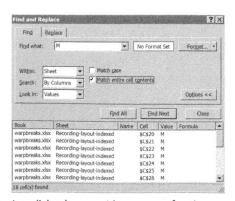

Figure 2.20 The *Find and Replace* dialog box provides a range of options.

You can choose to match the case of the target or to match only complete cell contents, or both. You can also choose to search for cells containing a specific format. This is very useful, as you'll see shortly.

If you click the *Find All* button you'll see a list of matches in the bottom pane of the dialog box (Figure 2.20). You can navigate to a specific match by clicking on the entry in the lower pane.

If you click on the *Replace* tab at the top you can choose what value to replace the target items with. You can also set the format of the replacement. You can choose to replace matches one at a time or to *Replace All*, which obviously does all in the list.

There can be unforeseen consequences of replacing items, and especially of doing a *Replace All* operation. If you have formulae that refer to the cells affected by the replacement the results will be different. This may be okay, or it may not! In the following exercise you can have a go at a *Find and Replace* and see the consequences of a global replacement operation.

Have a Go: Use Find and Replace

You'll need the *warpbreaks.xlsx* data file for this exercise. The file contains two worksheets: *Recording-layout* contains the basic data and *Recording-layout-indexed* includes extra index columns.

↘ Go to the website for support material.

1. Open the *warpbreaks.xlsx* file and select the *Recording-layout-indexed* worksheet. You can see the formulae by clicking on the cells.
2. Column C contains a predictor variable that has three levels; L, M and H. These are different tensions that two types of wool were subjected to. You will replace the values with Low, Med and High. Start by clicking the *Home > Find & Select > Find* button.
3. Click on the button labelled *Options >>* to open out additional options. If the button is labelled *Options <<* there is no need, as the extra options are displayed already.
4. In the *Find what* box type L.
5. Click the boxes labelled *Match case* and *Match entire cell contents* so that both are ticked.
6. Click in the *Look in* section to open up the options, select *Values*.
7. Now click *Find All*. The results are shown in the lower pane but you may need to use the mouse to make the box larger.
8. Click on the *Replace* tab at the top of the dialog box to bring up the replacement options.
9. In the *Replace with* box type Low. This will replace the L.
10. Now click *Replace All*. The matched values of L are replaced with Low and you'll see a box telling you that this was done 18 times. Click *OK* to close the box.

11. The *Find and Replace* dialog box remains open. Change the *Find what* item to M and the *Replace with* item to Med and click *Replace All*.

12. Finally replace the H items with High and close the *Replace and Find* dialog box (use the *Close* button).

You have now successfully replaced the short predictor variable with longer, and perhaps more meaningful, values. However, the formula in column G combined the variables *Wool* and *Tension*, so it has altered as well. If you want to keep the original label you'll need to modify the formula. (=B2 & LEFT(C2, 1) would do the trick).

This highlights the value of planning; try not to get to a situation where you have to alter items wholesale. This is easier said than done of course. There are occasions when you can overcome the problem, but this does depend on the formulae involved. In the following example you'll have a go at using conditional formatting to prevent a formula from updating incorrectly.

Have a Go: Use formatting in Find and Replace

You'll need the *mtcars.xlsx* data for this exercise. The file contains two worksheets, *Raw-data* and *Indexed-data*. The latter includes extra index variables.

↘ Go to the website for support material.

1. Open the *mtcars.xlsx* spreadsheet. Navigate to the *Indexed-data* worksheet. This contains additional index columns, made using various formulae.

2. Column A contains model names for the various cars for which you have data. You want to replace the Merc text with the full name Mercedes. Start by clicking the *Home > Find & Select > Find* button.

3. In the *Find what* box type Merc.

4. You want to check *Match case* but uncheck *Match entire cell contents*.

5. If you carry on you will replace Merc items in column Q, which has an index variable. Click on cell Q2; there is no need to close the *Find and Replace* dialog box. You'll see the formula looks for Merc* so even if you alter the original to Mercedes the formula will work. In addition, the value the cell returns will be Merc or Other. You could simply allow the replacement but there is a way to prevent it.

6. Click the header bar for column A to select that column; there is no need to close the *Find and Replace* dialog, keep it open. Now hit the *Home > B* button to make the cells **bold**.

7. Click in the *Find and Replace* dialog box to reactivate it. Now click the triangle by the *Format* button to bring up some options, select *Choose Format From Cell*.

8. The *Find and Replace* dialog box disappears and you can click on a cell to

pick up its formatting: click cell A1. The *Find and Replace* dialog should reappear.

9. Click in the *Look in* box and select *Values*.
10. Now click the *Replace* tab at the top. Type Mercedes in the *Replace with* box.
11. Click the *Find All* button and you should get a message at the foot of the box telling you that 7 items were matched. Make the match pane larger using the mouse to see the matches. Notice that the only items matched are in the *Model* column (column A).
12. Now click the *Replace All* button. You can now close the *Find and Replace* dialog box.

Now you have matched only the items in the column A, because they were **bold**. The formula in column Q remains unaffected by the change. You can leave column A bold or return it to normal.

As you can see from the preceding exercise, the formatting of cells can help in selective matching. Of course this trick does not work all the time but it may save you a bit of copy and paste. You can of course replace cells one by one; however this will not avoid the problem of formulae referring to cell contents that have altered.

Find and Replace can be useful but there may be consequences so think carefully.

2.2.2 Sorting data

Sorting your data can help you see samples together in blocks; this can help you spot errors and also get a better overview of what you have.

The *Sort & Filter* button on the *Home* menu is where you go to access the tools you need (Figure 2.21). You can also find the *Sort* tools from the *Data* menu (this is also true of other spreadsheet programs).

Figure 2.21 The *Sort* tools available from the Home menu in Excel 2010.

The simplest and quickest tools sort in ascending or descending order, the *ascending* button label says *Sort A to Z* or *Sort Smallest to Largest* depending on whether you have selected text or numeric data. The *descending* button is similar. If you simply want to sort your data using a single column you can follow these steps:

1. Click once anywhere in the column of data you want to sort.
2. Go to the *Home* menu on the ribbon.

3. Click the *Sort & Filter* button.
4. Click the button that sorts in the order you want.

The *Data* menu contains the same buttons for sorting.

Your data will be rearranged in order; note that Excel keeps your records together and your column headings intact. Note also that you do not need to select any cells.

The more columns you've got the more useful it is to be able to sort your data but sorting on a single column is of limited use. You need to be able to use more than one column to allow you to arrange your data in meaningful blocks. To do this you need the *Custom Sort* button on the *Home* menu (this is simply labelled *Sort* on the *Data* menu). The *Custom Sort* button opens a *Sort* dialog box that gives you various sorting options (Figure 2.22).

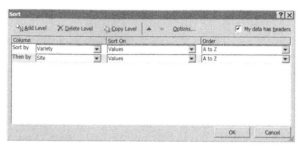

Figure 2.22 The *Sort* dialog box appears when you select a *Custom Sort*.

The *Sort* dialog allows you to specify several columns (Excel assumes you have headings, but you can turn this off) and to sort each one in various orders. The *Options* button also allows you to carry out a case-sensitive sort and to sort across rows rather than columns. In the following exercise you can have a go at sorting some data yourself.

Have a Go: Use Custom Sort on multiple data columns

You'll need the data file *Barley.xlsx* for this exercise. The file contains data from yields from several varieties of barley at different sites; each combination of variety and site were used in two years, 1931 and 1932. You want the worksheet *Barley*. The other worksheet, *Barley-Labelled* contains a version with extra label and index columns.

↘ Go to the website for support material.

1. Open the *Barley.xlsx* spreadsheet and go to the *Barley* worksheet.
2. Click once anywhere in the block of data, then click the *Home > Sort & Filter > Custom Sort* button. The *Sort* dialog box should open.
3. Note that the data appears highlighted in the background, except for the top row. Excel assumes the top row is column headers. Untick the *My data has headers* box and notice that the selection expands to include the top row. Retick the *My data has headers* box.

4. Now click the *Sort by* box and select *Year* as the column to sort on. Click the *Sort On* box and make sure *Values* is selected. Click the *Order* box and make sure you select *Smallest to Largest*.

5. Click the *Add Level* button to add a set of sort criteria. Click the *Then by* box and choose *Site* as the column to sort on. Set *Sort on* to *Values* and set *Order* to *A to Z*.

6. Click the *Add Level* button again. Set the criteria to *Variety, Values* and *Z to A*.

7. Click the *OK* button to close the *Sort* dialog box and apply the sort.

8. Click the *Home > Sort & Filter > Custom Sort* button once more to reopen the *Sort* dialog. You will see the sort criteria in place.

9. Click on the row label *Sort by* to select this sort criterion. Shift the criterion down by clicking on the triangle by the word *Options* at the top of the dialog box. Shift the row to the bottom and click the *OK* button to apply the new sort.

The *Custom Sort* dialog box gives you plenty of ways to rearrange your data. Notice that you can sort on cell or font colour! This can be useful if you want to highlight a particular level of a variable, or even entire rows that represent specific observations.

Note: Sorting independent blocks

If you highlight a block of cells you can sort them independently from the rest of the data. If you highlight cells from a block that does not include all data columns (that is, not complete records) then a warning box appears, allowing you to expand your selection.

Sorting in a special order

You will have noticed that when you sort text items you are limited to sorting from A–Z or from Z–A. The sorting is usually not case sensitive but you can click the *Options* button in the *Sort* dialog box (that is, *Home > Sort & Filter > Custom Sort > Options > Case sensitive*) to make the sort case-sensitive.

There may be occasions when you want to sort items in a particular order that is not alphabetical. For example, you might want to sort on months of the year and days of the week (these are predefined in most spreadsheets). It is possible to make a *custom list* that can be used in a *Sort* operation. In Excel 2010 you can do this from the *Sort* dialog. From the *Order* box in the *Sort* dialog you can select *Custom List*, which opens a new dialog box (Figure 2.23).

You can now define and select your custom list. In the following exercise you can make a custom list for use in sorting data for yourself.

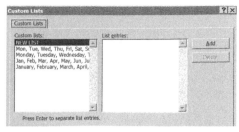

Figure 2.23 Custom lists can be defined and selected from the *Sort* dialog box, which opens the *Custom List* dialog box.

Have a Go: Make a custom list for sorting in a unique order

You'll need the *warpbreaks.xlsx* data file for this exercise. The file contains two worksheets: *Recording-layout* contains the basic data and *Recording-layout-indexed* includes extra index columns.

↘ Go to the website for support material.

1. Open the *warpbreaks.xlsx* spreadsheet and navigate to the *Recording-layout* worksheet. The *Tension* column contains a variable with three levels, L, M and H. These represent three tensions (Low, Medium and High) that two kinds of wool were subjected to on looms.
2. Click once in column C (Tension) and then sort (A–Z) using the *Home > Sort & Filter > Sort A to Z* button.
3. You'll see that the data are now arranged in order but you have an order of H, L, M. This is of course alphabetical. It would be more sensible to have L, M, H, which is a more natural representation of the actual tensions. Click again in the data and then use the *Home > Sort & Filter > Custom Sort* button to open the *Sort* dialog box.
4. Set *Sort by* to *Tension* and *Sort On* to *Values*. Now click the *Order* box and select *Custom List*, which will open the *Custom Lists* dialog box.
5. Click on the label that says *NEW LIST* and then click in the box headed *List entries*.
6. You need to enter the items in the order you want. Separate items with commas. So, type L, M, H. Then click the *Add* button. You should see your new list in the box on the left so make sure that it is selected (click on it) and click the *OK* button.
7. You should see a representation of your custom list in the *Order* box (L, M, H). You can now click *OK* to apply the sort.
8. Open the *Sort* dialog again and make a new custom list to sort the other way around, that is from High to Low. Click the *Order* box and select *Custom Lists*. Click *NEW LIST* and then type the items in the *List entries* box. This time, enter each item on a separate line, that is H, Enter, M, Enter and finally L. Make sure the new list is selected before clicking *OK*. Apply the sort with *OK*.

Once you have made a custom list it remains available in the *Custom Lists* dialog, even for other workbooks.

In the preceding exercise you made a custom list and then a second one to sort the opposite way around. You don't need to make the opposite list; once you've defined a list Excel allows you to use it forwards or backwards.

> **Note: Defining custom lists for sorting**
> In some spreadsheet programs you have to define custom lists for sorting before you apply the sort. Usually the definition can be made in the *Options* or *Preferences* dialog box.

2.2.3 Filtering data

You can use *Filters* to reduce the number of rows (records) of data that you see at any one time. You can think of this as an extension to using *Sort*. When you apply a sort you are rearranging your data in a particular order. *Filter* allows you to reduce the amount of data you display, so that you can focus on a smaller chunk of the dataset.

You can apply filters to your main dataset as well as any summary pivot table that you may produce (see Part 2). A basic filter simply hides those rows in your data that do not match the criteria you set. With an Advanced filter you can output the result to a new location, which can be useful if you want to share some of your data (Chapter 9) or to carry out an analysis on a portion of the dataset. There is also a special interactive filter, called a *Slicer Tool*, which you can use only on pivot tables; you'll see more of this when you look at exploring certain kinds of data (Chapters 6–8).

Basic filters
You start the filter process using the *Home > Sort & Filter* button. The *Data* menu also has a filter tool (Figure 2.24).

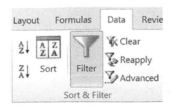

Figure 2.24 The *Data* menu on the ribbon has a Filter tool in the *Sort and Filter* section.

The basic filter process is quite straightforward:

1. Click once anywhere in the block of data.
2. Press the *Home > Sort & Filter > Filter* button (or use *Data > Filter*).
3. Column headings now display dropdown icons. You can click these to create filter criteria for each column.

The most basic filter is simply a selection tick-list, where you can hide (untick) or display (tick) levels of the variable from the list (Figure 2.25).

Figure 2.25 The *Filter* dropdown box allows various filter criteria to be designated.

If your chosen column contains text you'll see one option labelled *Text Filters*. Clicking on this brings up various options allowing you to modify the basic filter (Figure 2.26).

Figure 2.26 *Text Filter* options for text variables.

If your chosen column contains numeric data the option is labelled *Number Filters,* with similar options (Figure 2.27).

Figure 2.27 *Number Filter* options for numeric variables.

Using the filter options is fairly straightforward and largely a matter of common sense. You can also apply simple sorts to the filtered data. This is limited to ascending and descending orders so if you do want a different order, do the sorting before the filtering. In the following exercise you can have a go at applying filters for yourself.

Have a Go: Apply basic filters to a dataset

You'll need the data file *Barley.xlsx* for this exercise. The file contains data from yields from several varieties of barley at different sites; each combination of variety and site were used in two years, 1931 and 1932.

↘ Go to the website for support material.

1. Open the *Barley.xlsx* spreadsheet and go to the *Barley* worksheet.
2. Click once anywhere in the block of data then click the *Data > Filter* button. The header row now contains dropdown icons.
3. Click the dropdown icon in the *Site* header to bring up filter options for the *Site* variable.
4. Click the box labelled *(Select All)* to untick it. Then click the box by the Duluth item to display just that site. You could select more than one site, but leave it at just Duluth for now. Click *OK* to apply the filter. Note that the icon in the header row alters to resemble a funnel, showing that a filter is in force. You can also see that the row numbers are shown in blue, rather than black. This happens when there are cells hidden between the rows displayed. You're now showing only data from the Duluth site.
5. Now click in the *Variety* header to bring up the filter options for *Variety*. Click (or hover over) the *Text Filters* option and a new popup menu appears (see Figure 2.26). Click on the *Custom Filter* option. This opens a dialog box entitled *Custom AutoFilter*. Actually all the options open the same dialog! The difference is what appears in the first box.
6. Click in the first box, where it says *equals*, to bring up a list of options (similar to those of the previous dialog). Click the option called *contains* (you may have to scroll down).
7. Now click in the box to the right and type No. Then click *OK* to apply the filter. You are now displaying those varieties with "No" in the name. The *Site* filter is active as well so only Duluth rows are displayed.
8. Click in the *Yield* header to bring up the filter options for the *Yield* variable. Click the top option labelled *Sort Smallest to Largest*. The sort is completed immediately. The data has simply been rearranged, as you applied a sort, not a filter.
9. Click the *Data > Filter > Clear* button to clear all your filters. Note that this does not remove the filter icons: filtering is ready to go, you've just removed the filter criteria.
10. Click the dropdown icon in the *Site* header and select only the *Crookston* site.
11. Now click on the *Yield* header and bring up the options for the *Yield* filter. Click *Number Filters* and then select *Below Average*. The filter is applied

immediately. You are now showing only results that are below the (overall) average (mean) for the *Crookston* site only.

Try a few other options and see how you can view different blocks of the data. When you want to clear all the filter criteria click the *Data > Clear* button. To remove filtering totally click the *Data > Filter* button.

Once a filter is in place it can remain. If you add more data or alter the existing data the filter may not work correctly. To apply the filter again, click the *Data > Reapply* button.

Note: Filter a single column
If you select a single column you can apply a filter just to that one. In practice it is hardly worth the bother; you might as well filter all columns and apply a filter just to one.

There are many ways to filter your data and when combined with *Sort* operations you have a lot of control over how to display the data. The *Filter* operations simply alter what data is displayed; this allows you to use *Copy & Paste* to move a filtered block of data to a new location. Moving blocks of data to a new location can be useful for all sorts of reasons. In general you'd follow these steps:

1. Use *Data > Filter* to start the filter process.
2. Apply your filter criteria until you are happy with the block of data that results.
3. Select the filtered data using the mouse as normal; include the header row.
4. Copy the data to the clipboard using *Home > Copy*, right-click then *Copy* or press Ctrl+C.
5. Navigate to the location where you want to place the data.
6. Use *Home > Paste*, right-click or press Ctrl+V to paste the data into your chosen location. If you are pasting to a non-Office program Ctrl+V will usually work.

You can select criteria for any or all of the variables. The *Custom AutoFilter* dialog box allows you to select two criteria for any one variable (Figure 2.28).

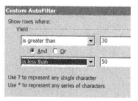

Figure 2.28 The *Custom AutoFilter* dialog box allows two criteria to be specified for a variable.

In most cases this is adequate but there are occasions when you may want more criteria. The *Advanced* filter option allows you to specify more criteria but the operation requires a bit more effort, as you'll see next.

Advanced filters

The *Advanced* filter option (on the *Data* ribbon) allows great flexibility and choice over the criteria but you use it in a different way to the basic filter. You must write the criteria into the worksheet as a separate block of cells, using headings to match the main data (Figure 2.29).

	A	B	C	D
1	Yield	Variety	Year	Site
2	>30	*No*	1931	Duluth
3	>30	*No*	1931	Morris
4				
5				
6	Yield	Variety	Year	Site
7	27	Manchuria	1931	University Farm
8	48.86667	Manchuria	1931	Waseca
9	27.43334	Manchuria	1931	Morris
10	39.93333	Manchuria	1931	Crookston
11	32.96667	Manchuria	1931	Grand Rapids
12	28.96667	Manchuria	1931	Duluth
13	43.06666	Glabron	1931	University Farm

Figure 2.29 The *Advanced* filter requires filter criteria to be specified as a separate block of cells in the worksheet.

You can choose to filter the output by hiding unwanted rows (as in the basic filter) or to send the results to a new location (Figure 2.30).

Figure 2.30 The *Advanced Filter* dialog box. You specify the main dataset (*List range*), the cells containing the criteria and optionally, the placement of the output.

Excel only lets you send the results to the active sheet. This makes it awkward to have the results placed on a new worksheet but it is possible, as you'll see shortly.

In Figure 2.29 you can see the criteria are specified as a block of cells above the main data. This makes it convenient to see what you are up to but potentially inconvenient for other purposes. You can still apply regular *Filter* and *Sort* operations and make pivot tables; the block of main data will be selected automatically as long as you click in it (and leave at least one blank row between the main data and the filter criteria). If you want to save the data (see Chapter 9) you'll have to delete the rows containing the filter criteria, either before or after saving.

It is possible to place the cells containing the filter criteria in a separate workbook. Your main data then remains unaffected and you can view the criteria and main data on the same screen with a bit of window juggling. In the examples that follow you'll see how to use *Advanced* filters using filter criteria in a separate workbook. If you decide to make space at the top of your dataset, that's fine.

You have two main options for dealing with the results of the *Advanced* filter:

- *Filter in place* – this is like the basic filter; rows are hidden from view to leave the filtered items on view.
- *Send results to a new location* – this makes a copy of the data, using only those rows that passed through the filter. You can only send the result to the active worksheet but you can work around that.

In either case you need to arrange the filter criteria as a separate block, using the same headings as the main dataset. The general way of proceeding is as follows:

1. Copy the data headings to your chosen location; either make space at the top of the main data or use a new worksheet.
2. Type the criteria you want to apply under the new headings you created in step 1.
3. Return to the main data and click in the range of cells.
4. Click the *Data > Advanced* button to bring up the *Advanced Filter* dialog.
5. Click in the *List range* box and select the main data (probably automatically selected).
6. Click in the *Criteria range* box then select the range of cells containing the filter criteria.
7. Choose *Filter the list, in-place* or *Copy to another location*. Select the location if you choose the latter option; this must be the active sheet.
8. Click *OK* to complete the filter operation.

Advanced filter criteria

The filter criteria are arranged in rows and columns, like the main data. To apply a criterion to a variable you type it in the appropriate column. If you have criteria in the same row they are treated together, that is, *criterion_1* AND *criterion_2*. If you have criteria on separate rows they are treated independently, that is, *criterion_1* OR *criterion_2*.

For numeric variables you can type a value as a criterion; it will be matched exactly. You can use other symbols such as the "greater than" symbol and so on but if you start with = Excel thinks you are typing a formula, so you'll need to wrap the expression in quotes. Table 2.8 shows most of the options available for numeric variables.

Table 2.8 Advanced filter expressions for numeric variables.

Expression	Meaning
10 =10 ="10"	Equals ten.
>10 =">10"	Greater than ten.
<10 ="<10"	Less than ten.

Continued

Continued

Expression	Meaning
>=10 ">=10"	Greater than or equal to ten.
<=10 =""<=10"	Less than or equal to ten.
<>10 ="<>10"	Not equal to ten.

You can see that for numeric variables you can avoid typing quotes as you never need to start an expression with an equals sign.

When it comes to text variables you can match text strings with additional flexibility: you can use wildcards, that is, a character that can be anything. If you type a text string (you don't need to start with =) it will be matched exactly. Think of it as being "equal to". You can type <>string to match everything except the *string*, that is, "not equal to". You can use a ? to match any single character. You can use the * to match any number of characters. If you want to match a ? or * then you use the ~ as an escape character, in other words ~? matches a question mark (the ~ turns the potential wildcard off).

To match the beginning of a word you can type the letter you want to match followed by an asterisk. To match anywhere within a string you place asterisks either side. Matching the end of a word is more difficult; you might think that preceding a word with an asterisk would do the trick but it doesn't (you match anywhere). You can see a list of the expressions in Table 2.9.

Table 2.9 Advanced filter expressions for text variables. All are case insensitive.

Expression	Meaning
Abc Abc*	Matches text "Abc" exactly, also matches if "Abc" is the start of the target.
*Abc *Abc*	Matches text "Abc" anywhere in the target (including the start).

Essentially you can search for whole strings, the start of strings and middle bits of strings (which includes starts and ends). You'll see how to match string endings shortly.

Filter in place

The simplest method of using the *Advanced* filter is to filter in place; in other words you mimic the basic filter and hide rows that don't match your criteria.

It is helpful to have your filter criteria in a separate worksheet. However, this means that you cannot see the criteria and data at the same time without some juggling. What you need to do is to make a new worksheet and arrange the windows so you can see both:

1. Create a new worksheet for the filter criteria: use the *Home > Insert > Insert Worksheet* button.
2. Right-click the sheet name in the Tab list at the bottom of the screen. Use *Rename* and type something meaningful. `Criteria` seems appropriate.
3. Return to the main data and select the header row. Copy it to the clipboard.
4. Navigate to the new worksheet and paste the header cells. Any spot will do but cell A1 is as good as any.
5. Press the Escape key to clear the paste operation. This clears the marching ants from the selection.
6. Click once in the *Criteria* worksheet you just pasted into and then use the *View > New Window* button. You now have a window with the original data and a window for the filter criteria.
7. Click the *View > Arrange All* button to bring up the *Arrange Windows* dialog box (Figure 2.31).
8. Click the radio button labelled *Horizontal* then tick the box that says *Windows of active workbook*. Click *OK* to arrange the windows on the screen.
9. The lower window will probably display the same worksheet as the top window so click once in the lower window and then click the Tab that contains your main data.
10. Resize the windows using the mouse. Remember that they act independently; you click once in a window to activate it.

Once you've got the windows arranged to your liking you can carry on and define the filter criteria.

Figure 2.31 The *View* menu allows you to make and arrange windows. The *Arrange Windows* dialog box appears when you click the *Arrange All* button.

The top right corner of each window has icons allowing you to minimize, maximize or close the window. If you want to make the main data window full size you can use the icon to maximize it and use the *Restore* button later to return to the two-window view (Figure 2.32).

Figure 2.32 The *Restore* button looks like two overlapping rectangles. The upper set of icons deals with the program window and the lower set with the worksheet window.

The easiest way to get a feel for the *Advanced* filter is to have a go for yourself. The following exercise uses the Barley data you used for the basic filter.

Have a Go: Use Advanced filters to filter in place

You'll need the data file *Barley.xlsx* for this exercise. The file contains data from yields from several varieties of barley at different sites; each combination of variety and site were used in two years, 1931 and 1932.

↘ Go to the website for support material.

1. Open the *Barley.xlsx* spreadsheet and go to the *Barley* worksheet.
2. Make a new worksheet called *Criteria* and copy the data header row from the main data to the new worksheet. Then arrange the windows so that you can see the filter criteria at the top and the main data at the bottom. You can use the steps shown earlier to help you do this.
3. Click in the top worksheet, *Criteria* to activate it. Then click in cell D2 and type `Duluth`. You are going to filter out all items except those for the Duluth site.
4. Now click in the lower worksheet, *Barley* to activate it. Make sure you click on a data cell (click once more if you are unsure).
5. Open the *Advanced Filter* dialog by using the *Data > Filter > Advanced* button.
6. The data in the Barley worksheet should be automatically selected (look for the marching ants). If it is not then click in the *List range* box and use the mouse to select the data. Now click in the *Criteria range* box.
7. Click in the top window to activate it, then highlight the criteria cells (include the header row). Now click *OK* to apply the filter. You'll see only those rows relating to the Duluth site.
8. Now click in the bottom window and then clear the filter using the *Data > Clear* button.
9. Return to the filter criteria and click in cell C2. Type the value 1931 in the cell. You are now going to filter the data to include records from 1931 and the Duluth site.
10. Now repeat steps 4–7 to apply the new filter.
11. Clear the filter (step 8) and return to the filter criteria. Click in cell A2 and type `>30`. You are going to view rows with a *Yield* greater than 30, as well as the other criteria.
12. Repeat steps 4–7 to apply the new filter.
13. Clear the filter again (step 8) and return to the filter criteria. Delete the >30 from the *Yield* column. Click in cell B2 and type `*No*`. Then repeat this in cell B3 underneath. Put 1931 as a value in cell C3 and in cell D3 type `Morris`.

14. Follow step 4–7 to apply the new filter; make sure that you select all rows of the filter criteria. You have now filtered data according to *Site* "Duluth" OR "Morris", for *Year* 1931. The *Variety* filter uses wildcards to match names containing "No".

Don't forget that things in the same row are combined as AND. Things in the same column are combined as OR.

Once you've set up your windows and the filter headings it is fairly easy to carry out the filter operations. If you want to copy the results to a new location simply select the result data (click once anywhere within it and press Ctrl+A) and then copy to the clipboard.

> **Note: Data selection**
> For most operations that require the selection of a block of data you don't actually need to select any data! When you click within a block of data Excel finds occupied cells and selects them automatically. Actually Excel searches around the current insertion point so even if you click adjacent to a block of occupied cells they will be found.

Matching the end of a text string

Matching the ending of a text string is not quite as easy as for other text matching. In Open Office you can use *Regular Expressions* to match endings but this is not so for Excel. (The wildcard syntax is also subtly different to Excel.)

You need to create a formula that returns a result of TRUE or FALSE. The TRUE results are displayed and the FALSE ones are hidden, thus applying the filter. You can use the EXACT function, which matches two strings.

```
EXACT(text_1, text_2)
```

If the two items are a match the result is TRUE, otherwise it is FALSE.

Your filter criteria must contain headings that match the data but you can add more headings and use these as filter criteria as long as they produce a TRUE or FALSE result. Actually you only need the headings you use to match the original. However, it is easier to use all the headings as they "line up" with the original data and let you see what you are doing more clearly.

If you wanted to match the ending of a text string you could use the RIGHT function. The general approach would be as follows:

1. Create your filter criteria in a new worksheet and arrange the filter criteria and data windows so you can see them one above the other.
2. Add an extra heading at the end of the criteria list; call it something meaningful, but short.

3. In the new column type a formula to match the rightmost end of your target from the data worksheet, with the text you want to match. You can use the mouse to help select the appropriate cells. Start with =EXACT("my text"), where "my text" is the text you want to match.

4. Type a comma and then continue the formula with RIGHT(cell, n). You can use the mouse to click the *cell* part in the main data window. Click the cell in the appropriate column (depending on what you want to match) and the first row. When you've done this type another comma and enter the *n* part. This will be a number, which should match the number of characters you typed in "my text".

5. You will need to close the parenthesis for the RIGHT function and add another to close out the EXACT function. You can then enter the formula.

6. You should see a result in the cell of TRUE or FALSE, depending on the contents of the first row of the data. Apply the *Advanced Filter* as usual but be sure to include the column with your new criterion as part of the *Criteria range*.

In Figure 2.33 you can see an example using the Barley data from the preceding exercise.

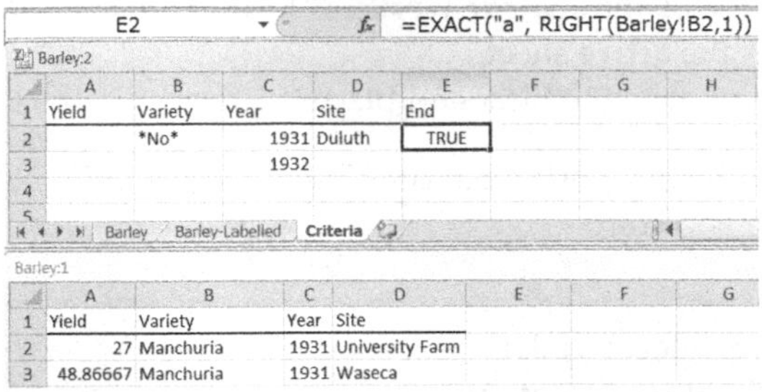

Figure 2.33 You can add extra filter criteria columns. Here the EXACT and RIGHT functions are used in combination to match the end of the Variety variable.

Once you have the hang of using formulae as part of your filtering, you can really unleash a lot of power. You could match items that are greater than the median for example, or that lie between the upper and lower quartiles. In any event the important thing is that your formula must return a result of TRUE or FALSE.

Filter to new worksheet

It is far easier to filter items in place than to get the results to go to a fresh worksheet. However, there may be occasions when you do not want to copy and paste your data and simply want it placed elsewhere.

Excel wants to place the results of your filter in the active worksheet; this is the one you were on when you activated the *Advanced* filter. The trick is to fool Excel into placing the results by being in the worksheet you want them to end up in when you click the OK button to apply the filter. Here are the steps you need to filter and place results in a new worksheet.

1. Make a new worksheet for the filter criteria and copy the data headers to the new worksheet. Name the worksheet *Criteria,* or some other meaningful name.
2. You don't have to bother arranging windows but it can be helpful to view criteria and data simultaneously when you are filling in the criteria.
3. Add the criteria to your *Criteria* worksheet.
4. Make a new worksheet to hold the filter results. Rename the worksheet *Output* or some other meaningful name.
5. Click in the *Output* worksheet and then activate the *Data* ribbon.
6. Click the *Advanced Filter* button; this will activate the *Advanced Filter* dialog box.
7. Since there is no data in the *Output* worksheet no data will be selected. Click the *List range* box and then highlight the data. You'll need to navigate to the main data worksheet and use the mouse.
8. Now click in the *Criteria range* box and select the cells containing your filter criteria, include the headers. You may have to click the *Criteria* worksheet to activate it before the mouse selection will work.
9. Click the *Copy to another location* radio button; this will enable the *Copy to* box.
10. Now click in the *Copy to* box and select the location for the results.
11. Click in cell A1 of the *Output* worksheet. Since this is where you started the process this is the active sheet.
12. Click *OK* to complete the filter and place the results in the *Output* worksheet.

Of course you can name the worksheets anything you like. It is best if the *Output* worksheet is blank before you start.

> **Tip: Selecting large blocks of data**
> If there is a large block of data to select you can click in a single cell and hit Ctrl I A on the keyboard; this will select all the data in a continuous block around the cell you first clicked in.

The filtering of data can help you extract chunks of your data, which you can use for various purposes such as checking for errors. Error checking is an important phase of your project and something that you'll explore further a little later (Chapter 3). Before that you'll look at a few other ways of rearranging your data.

2.3   REARRANGING YOUR DATA

Having your data in a particular order can help you manage them more effectively. Data in a spreadsheet can be sorted in two main ways:

- By row.
- By column.

Rearranging by row can be done using the *Sort* tools that you've met before (Section 2.2.2). It is also possible to use *Sort* to rearrange columns, but generally it is easier to do it manually.

2.3.1   Rearranging data by row

Rearranging your data by row can be helpful because it brings together data that are similar in some way. The *Sort* tools are found on both *Home* and *Data* menus (see Figures 2.21 and 2.24). You met these tools in Section 2.2.2 so here you'll see a brief summary.

When you carry out a *Sort* operation you have various options:

- Sort by values (numbers).
- Sort from smallest to largest or largest to smallest.
- Sort alphabetically from A to Z.
- Sort alphabetically from Z to A.
- Use multiple columns (Figure 2.22).
- Use a *Custom Sort List* (Figure 2.23).
- Use a specific block of data.

Generally speaking you'll find that using alphabetical sorting is most useful when you have categories of things and have predictor variables with discrete levels. You're most likely to be carrying out investigations involving association or differences between blocks of things. Custom sort lists fall into the alphabetic sorting category.

By contrast numeric sorting is most useful when your variables are all numeric (this is rather obvious). In these cases you are most likely to be looking for correlations; that is links between numeric variables.

Often your dataset will contain some columns with numbers and some with alphabetical entries. In such cases you may choose to apply a mixed sort, for example you may rearrange your data into sampling blocks and have each block arranged in numerical size order.

You can carry out sort operations quickly and easily apply different sorts as needed.

2.3.2   Rearranging data by column

Arranging your data by column is helpful for two reasons:

- For charting.
- For keeping track of data and index variables.

It is generally helpful to have your variables grouped. Most helpful would be to have the response variables next to one another and the predictor variables next to one another. Index variables can go in a separate block. Most often you'll have a single response variable and one or more predictor variables. You can place the response first or last. It seems to make more sense to have the response variables at the start, with the predictors following.

When Excel makes graphs it assumes data are in a particular order (you'll see this in Section 3.3). Excel assumes that the first column is for the *x*-axis, the predictor variable, and that subsequent columns are *y*-axis variables (response variables). For most scientific contexts this makes little sense!

It is usually best to build your charts (what Excel calls a graph) from a blank start, that is, an empty chart (one with no data). If you use named ranges you can make a chart quite quickly and as fast as highlighting data with a mouse. It does not matter in what order the columns are because you'll simply refer to them by name. In this case it makes more sense to have your response variables at the start, followed by the predictors and then the index variables.

If you need to rearrange your data you can do it in one of two ways:

- Click and drag – as the name implies, you use the mouse to move columns into place. This is the safe option as any cell references are updated as you move columns; so index variables point to the correct cells.
- Sort operation – sorting can be carried out on columns but you'll usually need to define a custom sort list. This option is useful if you have a large dataset without index variables.

Which method you choose depends upon your dataset and what you feel most comfortable with. Generally the larger the dataset the more helpful it is to use the *Sort* method. Hopefully as part of your planning process you set out your spreadsheet with the columns in a sensible order to begin with but there are often times when you need to do some tweaking.

Click and drag columns

The most direct way to rearrange your data columns is to drag them directly into place using the mouse. If you are moving a column to the end of the line you won't need to make space to accommodate the column. If you want to place the column in the middle of some other columns you'll need to make space.

Shifting columns around is fairly straightforward. Simply follow these steps:

1. Decide where your column needs to be. If this is where a data column already exists you'll need to make space. If the target location is already blank then go to step 4.
2. Click the column header where you want to place the column; the entire column will be selected.
3. Either right-click and select *Insert* or use the *Home > Insert* button (Figure 2.34). The highlighted column shifts to the right, opening a space to accept the column you will move into place.

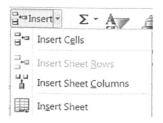

Figure 2.34 The *Home > Insert* button has options; the top option is carried out by default.

4. Click in the header of the column you want to move to select the entire column.
5. Hover over a selection border (a left or right edge is easiest) until the mouse cursor alters to a four-headed arrow.
6. Click and hold the mouse button. Then drag the column into the space you made in step 3. Release the mouse button to complete the operation.
7. Now click in the header of the empty column that was left behind (unless you want to move something into the space) to select the entire column.
8. Either right-click and select *Delete* or use the *Home > Delete* button (Figure 2.35).

Figure 2.35 The *Home > Delete* button has options; the top option is carried out by default.

Moving columns around like this is easy enough but can be tedious if you have many columns to shift. In such cases it can be easier to use a *Sort* operation, as you'll see next.

Note: Insert and Delete buttons
The *Home > Insert* and *Home > Delete* buttons insert and delete (as the names suggest) according to what is selected. If you select an entire column then the default will always be to insert (or delete) the column. If you click the triangle icon on these buttons you'll see additional options (see Figures 2.34 and 2.35). These allow you to insert (or delete) an entire column even if it is not selected.

Sort by column
There is a version of *Sort* that allows you to rearrange columns in your dataset. You can access the *Sort* tools from the *Home* menu (*Sort & Filter > Custom Sort* button, Figure 2.21) or from the *Data* menu (*Sort* button, Figure 2.24).

Note: Column sorting and index variables
If you have index variables using a column *Sort* operation will probably alter them. It is safer to use the mouse to click and drag columns, as the cell references will be altered along with the move.

In either case the *Sort* dialog box that appears (Figure 2.22) allows you to sort by rows (the default) or columns. You need to click the *Options* button, which will let you decide which *Orientation* to apply to the *Sort* (Figure 2.36).

Figure 2.36 The *Sort* dialog box has an *Options* button, allowing data to be sorted by row or column.

Select the *Sort left to right* option to apply the *Sort* to columns. You'll need to choose which row to sort on; *Row 1* would be most sensible, since it contains the column headings (variable names). You may be able to get away with a simple alphabetical sort but most likely you'll need to make a *Custom List* to arrange the columns in the order you want.

> **Note: Column names to sort**
>
> If you are sorting by column and creating a *Custom List* to use in the *Sort* there is no need to include all the columns. Name the ones you require to be sorted and anything not named explicitly will be sorted alphabetically after the columns you do name.

You met the *Custom List* for *Sort* operations previously (in Section 2.2.2) but as a brief reminder here are the steps you'll need to take to carry out a column sort:

1. Click once anywhere in the block of data.
2. Click the *Data > Sort* button to open the *Sort* dialog box.
3. Now click *Options* and select *Sort left to right* in the *Orientation* section. Click *OK* to set this option.
4. In the *Sort by* box use the triangle dropdown to select *Row 1*.
5. Make sure the *Sort on* box shows *Values*.
6. Click the dropdown icon in the *Order* box and select *Custom List*; this opens the *Custom Lists* dialog box.
7. Click the *NEW LISTS* item in the *Custom lists* section on the left.
8. Now click in the *List entries* box on the right and type the names of the columns that you want sorted. You can press Enter between items or use a comma. You should match the case of the column names.
9. Click *Add* to create the new *Custom List*. Now click *OK* to return to the *Sort* dialog box. You should not need to add any other items so click *OK* to complete the *Sort*.

Once you have a *Custom List* in place you can alter it by clicking on it in the *Custom lists* section as you make a new *Sort*. You can alter the entries as you like and click *Add* to complete the operation. You can also delete *Custom Lists* from this dialog (Figure 2.23).

Once you have made a *Custom List* you'll be able to see it in the dropdown box in the *Sort* dialog box. You'll also see the reverse sort (Figure 2.37).

Order
Yield, Variety, Site, Year ▼
A to Z
Z to A
Yield, Variety, Site, Year
Year, Site, Variety, Yield
Custom List...

Figure 2.37 Custom lists used in sorting appear with their reverse.

Note: Custom Lists in Open Office

If you use Open Office (or Libre Office) you must define your *Custom List* for sorting before you begin the *Sort* process. Use *Tools > Options* to get the options dialog box. Then open the *Calc* options and click on *Sort Lists*. You can now create and save your list.

You should be careful when sorting columns, especially when you have index columns that refer to other variables. If you alter the position of the variable columns you'll alter the index variables too. If you do have index columns then it is safest to use the click and drag method to rearrange columns, as the cell references are updated as you shift the columns.

2.4   EXERCISES

1. Of the following instances of the COUNTIF function, which one is NOT valid?

    A) =COUNTIF(A2:A99, A2)
    B) =COUNTIF(A2:A99, > 21)
    C) =COUNTIF(A2:A99, "A2")
    D) =COUNTIF(A2:A99, 21)
    E) =COUNTIF(A2:A99, "21")

2. The DATEVALUE function returns a number that corresponds to the Excel date; this is a value where January 1900 is 1 (for most versions of Excel). TRUE or FALSE?
3. When you need to select all the data in a worksheet you can use _____ as a shortcut from the keyboard rather than use the mouse.
4. Lookup tables can be useful to make _____ variables. The two main functions used are _____ and _____.
5. Open the *jackal.xlsx* data and navigate to the *Recording-layout* worksheet. How do you display records that are above the average mandible length?

↘ Go to the website for exercise files.

The answers to these exercises can be found in Appendix 1.

2.5 SUMMARY

Topic	Key points
Arranging columns	In general you should place your response variable in the first column followed by the predictor variables. You should place index variables and other information in subsequent columns.
Index variables	The COUNTIF function can be used to generate index values for variables by matching a criterion, which is usually the value of the target cell. The criterion must be in quotes, unless it is a plain number or cell reference, but it is not case sensitive.
Joining items	The & symbol can be used in an Excel formula to join items together. This can be helpful when making index variables.
The $ symbol	Cell references can be fixed by prefixing with the $; row and column references are independent so you can fix the row, column or both. When the cell formula is copied the fixed reference does not alter.
Dates	Cells can have a special date format, which can vary between countries. It is useful to maintain separate variables for data components for example, year, month and day. The DATE function can take the components and make an Excel date. Other functions can take an Excel date cell and extract a component, for example YEAR.
	The Julian day is a common method for representing the position of a date within a year. 1 January is day 1; each subsequent day increases the value by one.
Times	Excel has special formats for times; these work broadly like those for dates. The TIME function can take hours, minutes and seconds as separate items and construct an Excel time. Excel uses 24-hour time format.
Type of variable	There are various forms of variable: numeric (that is a continuous variable), categorical (number or text); binary; labels; replacements; abbreviations.
Text functions	Excel has various functions related to text, which can help you make grouping variables. The LEFT and RIGHT functions extract a number of elements from one end or the other of a target cell. Other functions are MID, FIND and LEN, which can be useful for making abbreviations for example.
	The ROMAN and ARABIC functions allow you to convert between regular numbers and Roman.
	The VLOOKUP and HLOOKUP functions allow you to replace one value with another using a reference table.
Paste Special	The *Paste* button gives several options, including *Paste Special*. These options allow you finer control over what is pasted from the clipboard. You can paste just values for example. You can even transpose a block of data so that the rows become columns and vice versa.

Topic	Key points
Notes	You can keep notes about variables, methodology, site information and so on, in various ways:
	Cell comments appear as popups when you hover over the cell with the mouse.
	In-cell notes can be typed as regular text. You can alter the cell wrapping and merge with adjacent cells to have the notes displayed in a more useful way.
	Text boxes are useful items that float above other worksheet elements. They are easily moved and resized.
	It is generally best to keep notes in a separate worksheet (labelled *Notes!*).
	Paper notes should be bound in some form. Loose sheets of paper tend to get lost. Try to lay out notebooks or recording sheets to mimic your planned dataset as far as possible.
Find & Replace	*Find and Replace* tools can help you make minor changes to your dataset. You can use *Replace All* for global changes. You can use conditional formatting to target changes.
Sorting	The Sort tools can help you reorder your data, using one or more variables as the sort key. This can group similar items together, which helps you to make changes and generally spot errors.
	You can make a *Custom List*, allowing you to specify sorting in a special order.
Filtering	The Filter tools allow you to display certain records in your dataset, according to various criteria. The regular filter hides unwanted data on a worksheet.
	The *Advanced* filter allows you to filter the results to a new worksheet (or workbook). You can use the EXACT function to make a special filter when matching cell contents.
Rearranging data	Rearranging data can be useful to help keep similar items together. Sort tools can rearrange the rows. Sort tools can be set to rearrange columns but this can be cumbersome to set up (since you'll need *Custom Lists*).
	You can rearrange columns by dragging them into a new position. Keep response and predictor variables together in blocks to make it easier when creating charts.

3

MANAGING YOUR DATA: CHECKING YOUR DATASET

An important, but often neglected, aspect of data management is error checking. You should not think that you need to wait until you've finished collecting and entering your data before error checking. Look at your data often and see how things are progressing. Spotting errors early can save you a whole heap of time and effort.

Some errors are fairly easy to spot and others are not. There are two sorts of error in data entry:

- Incorrect text entry.
- Incorrect numerical entry.

Incorrect text entries are mostly simple spelling mistakes (typos). This includes the addition of spaces at the end of entries, which are hard to spot. There are various ways to deal with text spelling errors, as you'll see shortly.

Errors with numerical data are far harder to deal with. Some values are obviously just wrong; an air temperature of 100°C is not likely (not on Earth anyhow) and should probably be 10. Some errors are less obvious. In these cases you may have to resort to graphical methods to spot oddities. Just because a datum is odd doesn't mean it is incorrect, but it is worth looking at more carefully. You'll look at these numerical oddities in Section 3.3.

Some errors can be avoided by using a validation process as part of the data entry. You can also extend this process to highlight odd data afterwards; you'll see this data validation in Section 3.2.

3.1 TYPOGRAPHICAL ERRORS

Spelling errors and general typos in text entries are inevitable and the more complicated your data labels are the more likely it is that these typos will creep in. You can minimize the occurrence of errors by validating data as it is entered (see Section 3.2), but you'll always need to look over your data. If you need to do some tinkering with the data entry it is best to make the changes early, when you have fewer data items to manipulate.

There are three main ways you can be a bit more systematic about error checking than simply looking over your data:

- Sorting – putting your data in various orders can help you spot oddities in spelling.
- Filtering – rearranging your data into logical chunks can help you spot mistakes.
- Pivot tables – rearranging your data in various ways can help you to spot errors.

3.1.1 Sort tools in error checking

Simply arranging your data in alphabetical order can help you spot potential mistakes. You can apply sorts in various ways, according to which variables you want to look over. Having similar entries next to one another helps you to check for errors, as the oddities tend to stand out.

If you want to keep the original sort order, that is, the data entry order, you should make sure that you have an index variable with the record number in each row. You can then sort on this column to restore the original order.

3.1.2 Pivot tables in error checking

The PivotTable tool was introduced in Section 3.2 as a way to help rearrange and visualize your data. You can also use pivot tables to help you check for errors, as you'll see now.

If you have a variable that has several different levels, there is a potential to misspell an entry. This kind of misspelling is particularly hard to spot if you have a lot of data records. For example if you are collecting data on different species you'll be managing many different names. Similarly you may have data collected from various sites. You can use the *Sort* functions to help you rearrange your data in order, using the variable in question. This will help you spot odd entries that are misspelled. However, this still involves you having to scan lots of data. This is where you can use the PivotTable tool to help you. If you make a pivot table using the variable as the *Row Labels*, you'll end up with a list of levels of that variable. You've now reduced the number of items you need to check. If you see two entries that are very similar, one may well be a misspelled entry. You can then use the *Find and Replace* tools (Section 2.2.1) to correct the mistake(s).

To check a dataset for typographical errors you'll need to follow a few basic steps:

1. Click in your main data and use the *Insert >PivotTable* button.
2. Choose to place the pivot table in a new worksheet.
3. From the *PivotTable Field List* drag the variable you want to check from the list at the top to the *Row Labels* section.
4. You should now see a list of the levels of the variable in your part-formed pivot table. Scan the list for entries that appear similar.
5. If you see two similar entries (they will most likely be next to one another) one is probably a misspelling of the other. Click the incorrect entry and copy the text in the formula bar to the clipboard.
6. Return to the main data worksheet and click the *Home > Find & Select > Find* button.
7. Paste the misspelled entry into *Find what*.

8. Enter the correct text in *Replace with*.
9. Click on *Replace* to correct the entry.
10. Use the *Find Next* and *Replace* buttons to correct all similar mistakes.
11. Return to the pivot table. Click in the data to bring up the *PivotTable Tools* menu.
12. Use the *Options > Refresh* button to redo the pivot table; the error you just corrected should disappear.
13. You can now repeat steps 4–12 until you have found and corrected the errors.

Another kind of spelling error that is very hard to detect by simple inspection is where an entered text variable has an additional space after the entry. You won't see the extra space but the computer will, and treats the record containing the space as a separate entity. The pivot table can help you spot this error too as you'll see two entries that appear identical. You can then use *Find and Replace* to find the variable with the extra space and replace it with an entry with the space removed. It is also possible to use the *Advanced Filter* (Section 2.2.3) to filter records containing a space at the end, which you can then edit manually.

In the following exercise you can have some practice at checking a dataset for typographical errors.

Have a Go: Check some data for typographical errors

You'll need the file *Error Checking.xlsx* for this exercise. There are two worksheets, *Original* and *Fixed*. The latter contains the completed version (there are errors other than typographical).

↘ Go to the website for support material.

The data represent abundances of various plant species in quadrats (sampling units) at two sites. Column A contains the species names. The next column has common names. Columns C and D contain the quadrat number (1–5) and the Domin score (an abundance value in the range 1–10). The final column gives the site name.

1. Open the spreadsheet and go to the *Original* worksheet. This contains the data with the errors. The *Species* column is what you'll be checking. Click once anywhere in the block of data then start the pivot table process with the *Insert >PivotTable* button.
2. The data should be selected automatically so click *OK* to make the pivot table in a new worksheet. Drag the *Species* field to the *Row Labels* box of the *PivotTable Field List* task pane.
3. The pivot table is only part-formed but it's sufficient to check the *Species* variable. Look at the first three entries, they are very similar. In fact the top entry is misspelled; *milefolium* should have double "l": *millefolium*. Click in the cell that contains the misspelled entry.
4. Copy the cell to the clipboard (Ctrl+C is easiest).

5. Press the Esc key on the keyboard to exit the formula bar then click the *Original* worksheet to bring up the data.
6. Now click the *Home > Find & Select > Find* button to bring up the *Find and Replace* dialog box.
7. Click in the *Find what* box and paste the misspelled entry with Ctrl+V.
8. Now click the *Replace* tab and in the *Replace with* box paste again. Correct the entry (add the extra 1).
9. Click *Find Next* and then *Replace* to correct the first incorrectly spelt entry. Try this again and you'll discover that there was only one instance of this particular error. Click the *Close* button to exit the *Find and Replace* dialog and return to the main data.
10. Return to the pivot table worksheet. Click once in the table to activate the *PivotTable Tools* menu. Now click *Options > Refresh* and the misspelled entry should disappear.
11. The top two entries look identical, *Achillea millefolium*. In fact one of them has a trailing space, which you cannot see. Click in one entry and then the formula bar to see which one it is. You'll find that the second entry is the one with the extra space (the cursor shows where the additional space is, so click at the rightmost end of the formula bar).
12. Click the incorrect entry and then select it in the formula bar; include the extra space. Copy the entry, with its additional space, to the clipboard.
13. Now follow steps 6–10 to *Find and Replace* the incorrect entries for *Achillea millefolium*.

You can now repeat the process for all the potential errors in the species names. You can see what the correct names are by looking at the *Fixed* worksheet (make a pivot table of species from the *Fixed* worksheet).

Once you've been over the data and checked the typographical errors for one variable you can check another using the same pivot table. Simply replace the variable you've just checked with the one you want to check in the *Row Labels* box. As long as you use the *Refresh* button after making alterations you won't have to make the pivot table afresh.

Tip: Reselecting pivot table data
If you've lost track of your edits and want to redo a pivot table you don't have to start afresh. Click once in the existing pivot table then use the *PivotTable Tools > Options > Change Data Source* button. Delete anything in the *Table/Range* box. Then click once in the block of data and then use Ctrl+A. This selects all the data. Click *OK* to refresh the pivot table with the new data selection.

Pivot tables are useful to help check your data for errors but you can also use *Filter* tools.

3.1.3 Filter tools in error checking

If you apply a regular *Filter* to a column you want to check, you can see the list of entries (Figure 3.1). Click the *Select All* label to deselect everything. Then scroll down the list, if you see a potential error you can select it to filter the column, then correct the error manually.

Figure 3.1 A regular filter can help spot typographical errors by showing similar entries.

Once you've corrected the error(s) you can click the filter icon in the column heading and carry on looking down the list.

You can also use the *Advanced Filter* tool. It is especially helpful in searching for entries with additional spaces at the end. You need to:

1. Make a new sheet containing the filter criteria, as you saw in Section 2.2.3.
2. Add a new heading and column for the special criterion.
3. Then enter a formula, you'll need to use EXACT and RIGHT functions as you saw earlier to search for a single space. Point to the first entry of the variable you want to check; the filter will work down the column.
4. Return to the main data worksheet and click once in the block of data.
5. Bring up the *Advanced Filter* and filter in-place, making sure you select the special criterion you just created.
6. The data are now filtered and any visible entries must contain extra spaces at the end. You can now edit the entries.
7. Clear the filter once you have finished to restore to view the entire dataset.

It is of course possible that the entries with the additional spaces have other errors!

Typos are tedious to spot and it would be nice if you could reduce the potential number of errors right at the data entry stage. It is possible to use *Validation* tools to help you do exactly this. These *Validation* tools also allow you to check data that you've already entered, giving you another weapon in the fight against errors.

3.2 VALIDATING ENTRIES

It is possible to restrict data entry to certain criteria, thus reducing possible errors. For example, if you are entering abundance on a DAFOR scale you can restrict the contents of the abundance variable to include only appropriate values (the letters DAFOR and maybe "N" for not present). If you are entering numerical data that are percentages you can restrict entries to 100% maximum.

The validation tools are found in the *Data* menu (Figure 3.2).

Figure 3.2 Data validation is carried out using tools in the *Data* menu.

There are two main ways you can implement *Data Validation*:

- *During entry* – this reduces the chances of errors, as you restrict what can be entered into the data sheet. You set up criteria, which are matched as the user enters data. There are options for how you deal with possible violations.
- *After entry* – this is a way to check already entered data for possible errors. You set up entry criteria, allowing possible errors to be flagged in the main dataset.

In general the larger the dataset the more helpful it is to have your data validation carried out at the entry stage.

3.2.1 Validation during entry

Error checking should start right at the beginning with data entry; if you can minimize input errors you'll have an easier time of things later. When you are designing your data input worksheets you can use the *Data Validation* tool to guide the input of valid data.

When you click the *Data > Data Validation* button you see the *Data Validation* dialog box, which lets you alter three settings (Figure 3.3):

- *Input criteria* – this is where you set the allowable entry criteria.
- *Input message* – this allows you to guide the user, helping them to enter data accurately.
- *Error alert* – this allows you to set what happens if data does not match the entry criteria.

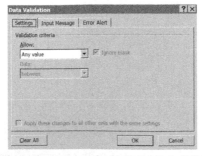

Figure 3.3 Data validation has three elements, as shown in the *Data Validation* dialog.

These three settings each have their own options, as you'll see now.

Input criteria

The input criteria you set will determine what is allowable in the cells you've selected. First of all you select the kind of data you want to permit, using the *Allow* field in the *Data Validation > Settings* dialog box (Figure 3.3). There are eight options:

- *Any value* – this is essentially a way to allow anything, and is what you set to remove a previously set criterion (or you can use the *Clear All* button).
- *Whole number* – allow integer values only; you can then set further matching criteria for the values.
- *Decimal* – allow any kind of number.
- *List* – allow text values that match a lookup table. You can show the options as a dropdown list, allowing the user to select with the mouse.
- *Date* – allow date entries; you then choose a range for the entries.
- *Time* – allow time entries.
- *Text length* – allow any kind of entry but restrict the length of the entry.
- *Custom* – this allows other sorts of entry; you type (or select) a formula to use to create the criterion.

Most of these options allow you to select the way the entry is matched; you can select to match between two values for example (Figure 3.4).

Figure 3.4 Entry matching can be refined using various options.

If you select the *List* type of entry you will need to define the values you are prepared to accept. These should be in a lookup table, preferably in a new worksheet. You simply type the possible entries in a column. Using a header is useful, to remind you what the lookup is for, but you do not select the header when you are choosing the *Source* for the *List* criterion (Figure 3.5).

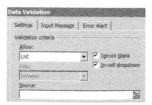

Figure 3.5 List entries use a lookup table; you enter the cells containing the possible entries as the source.

You can choose to have the list presented to the user as a dropdown; the user will see a dropdown icon when they select the cell. If the user clicks the icon a list appears that allows the user to select an appropriate item (Figure 3.6).

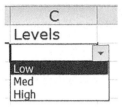

Figure 3.6 Data cells that have a list criterion as a dropdown display the options as a dropdown list.

Once you've worked out what kind of entry you are prepared to accept you can move on to guide the user by setting on screen prompts.

Input message

You can choose to display a helpful message when a cell is selected. The message can contain a title and a main body (Figure 3.7).

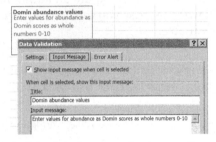

Figure 3.7 Input messages can be displayed to users when input cells are selected.

Ideally you want the message to be short, but helpful. The message will only be displayed when the cell is selected. If you click the *Clear All* button you'll remove the input message as well as any *Input Criteria* or *Error Alert* you've defined.

These messages can be useful even if you expect to be the person entering data.

Error Alert

You can choose what happens when invalid data are entered into the cells you are validating (Figure 3.8).

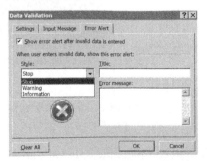

Figure 3.8 You can choose what happens when invalid data are entered as part of data validation.

There are three options; each allows you to choose a title and text for the warning box that will appear. The difference between the options is what happens when an error is detected:

- *Stop* – will not allow an incorrect entry; the user can retry.
- *Warning* – will allow an incorrect entry; the user is invited to accept the current entry or not.
- *Information* – allows incorrect entries with a simple message to say that the entry does not match.

So, the *Stop* type is most strict and the *Information* type is the least; which one you choose will depend upon the circumstances. As a general rule the *Warning* type is most flexible but where you have certain fixed levels of a variable, such as site name, you may want to be more rigid.

Numeric variables are simplest to validate, as you can simply determine the numeric range you want to include. Text variables require an additional stage, which you can practise shortly. In the following exercise you can have a go at setting up some validation criteria for a numeric variable.

Have a Go: Set up data entry validation for a numeric variable

You do not need anything except Excel for this exercise, as you'll make a workbook from scratch.

↘ There is a completed version of the exercise, *Validation.xlsx*, available on the companion website.

1. Open Excel and make a new workbook, rename the sheet to *Entry sheet*. Type a heading Domin in Cell A1.
2. Column A will contain Domin scores, that is, numeric (integer) values from 0 to 10. Click in the column heading, to select the entire column. Then click the *Data > Data Validation* button to bring up the *Data Validation* dialog.
3. Make sure you are in the *Settings* tab. Click in the *Allow* field to bring up a list of criteria types and select *Whole number*.
4. Click the *Data* box to bring up the options. Domin scores need to be in the range 0–10, so select *between* and then enter the values 0 and 10 in the appropriate fields (labelled *Minimum* and *Maximum*). You could try *less than or equal to* 10 but this would also permit negative values.
5. Go to the *Input Message* tab to set up a user guidance message. In the *Title* section type Domin abundance scale. In the *Input message* section type Whole numbers from 0 to 10.
6. Go to the *Error Alert* tab to set up the action on errors. Click the *Style* box and select the *Warning* option. In the *Title* box type Domin abundance and in the *Error message* box type Domin scores are whole numbers between 0 and 10.
7. Click the *OK* button to complete the data validation criteria for column A, *Domin*.

8. Because you selected the entire column the header label is also validated. Ignore that for now, as you can remove validation criteria from the entire header row at a later stage.

Now you can enter some values into the cells of the *Domin* column. See what happens if you enter inappropriate values. Try altering the *Error Alert* style and see how the response alters.

Numeric variables are generally easier to deal with than text variables. In the preceding exercise your entries could only be in the range 0–10, which was easy to manage. At other times you will not be sure of the range to expect. You could omit validation for the variable or put in your best estimates for the range and set the *Error Alert* style to the least strict *Information* setting.

Tip: Removing validation criteria from the header row
It is easier to select entire columns when making validation criteria. The means you end up with validation applied to headings. To remove the criteria, select the entire row, and then click the *Data > Data Validation* button. You may get a warning message because adjacent cells have validation criteria. If you are asked if you want to expand your selection, select *No*.

If you are using text variables then you'll need to define the range of options in a lookup table. In the following exercise you can have a go at this using free-form or dropdown styles of list.

Have a Go: Set up data entry validation for text variables
You do not need anything except Excel for this exercise, as you'll make a workbook from scratch.

↘ There is a completed version of the exercise, *Validation.xlsx*, available with the data on the companion website.

1. Open Excel and make a new workbook. Rename Sheet1 *Entry sheet*. Rename Sheet2 *Lookup sheet*. If you have only a single worksheet then add a new one and rename it.
2. Return to the *Entry sheet* worksheet. Type a heading in cell B1, DAFOR will do because you are going to make a text variable containing an ordinal abundance scale.
3. Go to the *Lookup sheet* worksheet that you made in step 1. In cell A1 type a heading Abund to remind you what the column is for. In the column cells underneath type D, A, F, O, R and N to represent the levels of the scale ("N" is for "not present"). The items can be in any order.

4. Return to the *Entry sheet* worksheet and select column B with the mouse. Now click the *Data > Data Validation* button to bring up the *Data Validation* dialog.
5. Make sure you are in the *Settings* tab. Click in the *Allow* box and select *List* from the options.
6. Click the box labelled *In-cell dropdown* to uncheck it.
7. Click in the *Source* box. Click on the *Lookup sheet* tab to bring up the lookup table, and then select the cells containing the entries (not the heading).
8. Click the *Input Message* tab then enter a helpful message: in the *Input message* section type Abundance as DAFOR or N.
9. Now click the *Error Alert* tab and select the *Stop* style. Then click *OK* to apply the validation to the *DAFOR* column.
10. Type some values into the *DAFOR* column. Try altering the *Error Alert* style.
11. Go to cell C1 in the *Entry sheet* worksheet and type a heading Tension, as a label for a text variable with set levels.
12. Now go to the *Lookup sheet* workbook and in cell B1 type Tension as a heading.
13. Start in cell B2 and type the levels of the Tension variable that are acceptable, Low, Med and High will be appropriate.
14. Return to the *Entry sheet* worksheet and select column C with the mouse. Then click *Data > Data Validation*, to bring up the *Data Validation* dialog.
15. Make sure you are in the *Settings* tab. Choose the *List* option in the *Allow* box and make sure the box labelled *In-cell dropdown* is ticked.
16. Click in the *Source* box and then select the three text items from your lookup table in the *Lookup sheet* worksheet. Use the other tabs to select sensible options for the *Input Message* and *Error Alert*.
17. Click *OK* to apply the validation criteria. You can now enter values in the *Tension* column. Clicking in a cell in this column brings up a dropdown icon, allowing you to select an option. You can still type something else; how this is treated is determined by the level you set the *Message Alert* to.

Try altering the settings and see if you can bypass the errors (stringent validation rules make it harder to bypass).

↘ Go to the website for support material.

When you've set validation for various columns you will have the validation criteria applied to the column headings as well as the data cells. Simply select the entire header row and use the *Data > Data Validation* button to obtain the *Data Validation* dialog. Then use *Clear All*.

Validating data as you enter is generally preferable to doing it later but inevitably you'll have situations where this is unavoidable.

> **Tip: Adding items to validation list lookup tables**
> If you've set up a *List* validation for a column you will have used a lookup table to hold the valid entries. If you want to add extra items insert the new items in the middle, not at the end). Use the *Home > Insert* button to insert a blank cell and shift the other items down. This will automatically update the lookup references in the validation.

3.2.2 Validation after data entry

You can use the *Data Validation* tools to check your data after they have already been entered. You set up the criteria using the same approach as you've already seen, using the *Data Validation* dialog via *Data > Data Validation*. The difference with validating data after entry is that you use the *Data > Data Validation > Circle Invalid Data* button (Figure 3.2) to highlight entries that do not match your entry criteria. Once you've highlighted the errors, you can edit them as appropriate. The *Clear Validation Circles* button acts like a reset button and clears the highlights (which are red circles around the errant cells).

The general way to proceed is as follows:

1. Select a column and then use *Data > Data Validation* to set up criteria to check against.
2. Select more columns as you need so that all the variables you want to check have validation criteria established.
3. Select the header row and use *Data > Data Validation* to bring up the *Data Validation* dialog. Clear the criteria for the header row with the *Clear All* button.
4. Highlight invalid data (that is, entries that do not match your criteria) with the *Data > Data Validation > Circle Invalid Data* button. It can take a few moments for Excel to perform this task if you have a lot of data, so be patient.
5. Invalid entries are circled in red. Check the entries and edit as required.
6. Clear all marks using the *Data > Data Validation > Clear Validation Circles* button.

If you have a large dataset you might want to check errors using one variable at a time, since the process can be time-consuming.

In the following exercise you can have a go at post-entry data validation for yourself.

> **Have a Go: Carry out post-entry data validation**
> You will need the *Error Checking.xlsx* spreadsheet for this exercise. The file has two worksheets, one called *Original* containing data with various errors. The other worksheet is labelled *Fixed*, and contains the data with the errors fixed.
>
> ↘ Go to the website for support material.
>
> The first two columns contain plant species names, in scientific binomial and common name form. The third column contains the quadrat number (*Qu*); these

values should be in the range 1–5. The next column gives abundance using the *Domin* ordinal scale; values here should be in the range 1–10. The final column gives a site name; there are only two, *upper* and *lower*. Each site has been surveyed five times. Only species that were found in a quadrat were recorded, so the *Domin* scores should all be > 0.

1. Open the *Error Checking.xlsx* spreadsheet; make sure you are in the *Original* worksheet. You are going to check the *Qu* and *Domin* variables. Click in the header bar for column C to select the entire *Qu* column.
2. Now click the *Data > Data Validation* button to display the *Data Validation* dialog.
3. Make sure you are in the *Settings* tab. Click in the *Allow* box and select *Whole number* and then set the *Data* section to *between*. Set the *Minimum* to 1 and the *Maximum* to 5.
4. You do not need to set anything for the *Input Message* or *Error Alert* sections, simply click *OK* to apply the validation criteria.
5. Now click in the header bar for column D to select the entire *Domin* column.
6. Repeat steps 2–4 but set the *Maximum* allowable value to 10.
7. Click in the row bar to select all of row 1 (the column headings). Click the *Data > Data Validation* button. You should get a message informing you that the selection contains various validations. Click *OK* to remove the validations, the *Data Validation* dialog should now be displayed. You can see that the *Allow* box says *Any value*, confirming that the validation criteria have been removed. Click *OK* to return to the data.
8. Click once anywhere in the data to clear your row selection. Now run the validation by using the *Data > Data Validation > Circle Invalid Data* button. It can take a few moments to complete the operation, so be patient.
9. Now you can look through the data for the red circles that highlight the potential errors. You can then edit as required; as entries are corrected the validation marks should disappear.

You'll see that in the *Qu* column there are a couple of invalid entries. The quadrat number is recorded as 6 when it should be 5. In the *Domin* column there is an entry of 11; this should be a 1. There is also an entry of 7. This appears to be in the 1–10 range but if you click the cell you'll see that it is actually 6.5. We expect integer values, so the entry is flagged as invalid. You can change it to 6.

In the preceding exercise you saw how to highlight potential errors. Correcting spelling mistakes is fairly easy but correcting numerical errors can be harder. In the exercise the *Festuca ovina* species was entered as found in quadrat 6 but this should have been 5. This was an easy mistake to spot, since quadrats 1 to 4 were all represented but 5 was missing. However, not all species were found in all quadrats. You may have to return to your original notes to work out what the correct entries should be.

Similarly, the *Domin* score of 11 is clearly incorrect but should it be 10, 1 or something entirely different? Looking at the general abundance of the species, which is low, seems to indicate that a value of 1 is more likely than a high value of 10. The entry of 6.5 also requires a judgment call.

You can use some of the other tools you've already seen to help you focus on the errors you spot during validation. For example, in the preceding exercise you could use a *Filter* to view the *Festuca ovina* species. You'd see quickly that the 6 entry should be a 5. A pivot table could also help view the situation: you'd need *Column Labels* set to *Site* and *Qu*, *Row Labels* set to *Species* and *Values* set to *Domin*. Once you know where the potential error is you can explore it in more detail to help you decide what the correct entry might be.

The numerical entries you've seen here are relatively easy to check, since there are fixed ranges for the entries. Other kinds of numerical data are not so restricted and checking these data is somewhat harder, as you'll see next.

3.3   NUMERICAL ERRORS

Numerical errors are usually the hardest to spot, especially if the variable in question has not got a particular fixed range. Usually your sample has a range of values that lie within the absolute boundaries of possible values. You are not looking for absolute errors but for entries that look odd.

Most data are variable: if you collect repeat measurements each one is slightly different. Usually you report your results using a *summary statistic*, such as an average (e.g. as a *mean* or *median*). You'll see how to summarize your data using summary statistics later (in Part 2). The observations lie around the average, with some larger and some smaller values. If most of the observations lie close to the average your data exhibits lower variability than a sample where the observations are more spread out, away from the average. The variability itself can be important and there are quantities such as *standard deviation* and *inter-quartile range*, which are measurements of variability. The variability between samples is used in most statistical analyses to help inform you of potential differences between the samples. Similarly when you are looking for links between variables, in correlation and regression, the variability of data is important in helping to decide if the link is statistically significant.

The spread of the data itself produces a pattern; this is called the *distribution* of the data. The classic case is called *normal distribution* (also called *parametric*) and shows a distinct pattern where most data lie close to the middle (the mean) and there are fewer data items as you get further away (from the mean). Other patterns are generally described as *non-parametric*. The *distribution* of the data is important in determining what kind of analytical test to use. You'll see how to visualize the patterns in your data (the data *distribution*) later (Part 2); for the moment you will focus on the error checking aspect.

There are two scenarios you'll need to consider; each requires a different approach to checking for odd data:

- Comparisons between samples.
- Correlations and regressions.

When you are exploring differences between samples, you need to look at the data on a sample-by-sample basis. If you were to represent the scenario graphically you'd use *bar charts* and similar graphs.

By contrast when you are looking for correlations or carrying out regression you are comparing two (or more) numeric variables (although there are other sorts of regression). These cases would lend themselves to visualization with *scatter plots*. You may want to break the data down by other variables and use sample groups but in any event you're still going to need a scatter-plot visualization.

You can use *Filter* and *Sort* tools to help you arrange and rearrange your data into chunks. The *PivotTable* tool also lets you reorganize your data in various ways. These tools allow you to scan the data and possibly spot those that appear odd. Just because a datum looks odd does not mean that it is wrong. Data are variable and having one or two that are way out could simply indicate that you haven't collected enough data. The point of the error highlighting is to give you the opportunity to differentiate between genuine errors and merely unusual entries.

Although *Filter* tools and the like are helpful, visual methods usually allow you to scan for oddities more easily. In the following sections you'll see how to explore the data in the two main scenarios.

3.3.1 Errors in sample data

If you took the values of a variable and arranged them along a single axis you'd see the points in a simple manner (Figure 3.9).

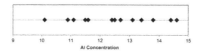

Figure 3.9 Sample observations arranged along a single axis form a simple dot chart.

You'd expect to see the points arranged with some variability: they would be spread out and not on top of one another (although around the average there might well be some overlap). The more spread out the points are the greater the variability. If you look at the points in Figure 3.9 you can see some variability but there are no points that seem way out. This kind of chart is called a *dot chart* (sometimes called a *strip chart*), and it is particularly useful for exploring your data when you have items arranged in samples.

A dot chart is essentially a scatter plot with only one axis. You can make scatter plots easily using the *Charts* section of the *Insert* ribbon. You can set the y-data to a single arbitrary value, such as 1, for a single sample. If you want to present data from more than one sample you give each sample its own y-value so that when plotted, each sample has its own level (Figure 3.10).

A disadvantage of the scatter plot layout in Excel is that you cannot label the samples; each occupies a level on the y-axis and has a purely numeric label. One way around this is to use a *line plot*, which allows you to label the samples clearly. There are disadvantages to the line plot, as you will see later, but these are relatively minor.

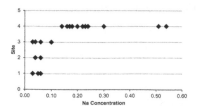

Figure 3.10 Multiple samples can be arranged in a dot chart; each sample occupies a level on the y-axis.

You can make dot charts in two ways:

- Use the raw data and the *Charts* section of the *Insert* ribbon.
- Use a pivot chart created from pivot table data.

If you use the raw data you can make scatter plots, whereas if you use a pivot chart you can make line plots. A pivot chart gives you a good deal of flexibility, because it is tied to a pivot table. If you alter the pivot table the chart updates to reflect the changes.

There is work to be done for either approach; which approach you take depends on the data you have. In the following sections you'll see both approaches.

Dot charts from raw data

From the raw data it is easier to construct scatter plots to use as your dot charts. The samples end up with simple numeric labels but you can easily produce a legend to inform readers which sample is which. In general you're going to use the dot charts for your own error checking and not for publication, so you can accept an imperfect solution.

To make a scatter plot you use the *Insert > Scatter* button. Before you get to that stage though, you'll need to prepare the data. The values you'll plot will be entered already for one axis; for a horizontal plot these x-values will already be in your data columns. The label axis will need some sorting out; you will have to make an index that contains a unique value for each sample. The value will have to be numeric so if you have site names for example, you'll need to make a new column with a number for each column. A lookup table can be very useful for this purpose. Once you have done that it is easy enough to construct the scatter plot.

The general running order to make a dot chart is as follows:

1. Decide how the samples are to be split; if you have a single predictor variable this is simple – there will be one value for each level of the variable. If you have two predictor variables you may want to make more than one index.
2. Make a new worksheet and create a lookup table so that you can convert the text-based levels of your predictor variables into a numeric value.
3. Return to the main data worksheet and use VLOOKUP to make a new column variable.
4. Click once in a location away from any data (at least one cell gap); Excel

searches around the current selection for any data cells, which it uses for the chart. Inevitably the chart is wrong so better to have no data and select it yourself.

5. Use the *Insert > Scatter* button to create a blank scatter plot, which you will populate shortly, choose an option for points only if given the choice.

6. Click once in the chart to enable the *Chart Tools* menu. Then use the *Design > Select Data* button to bring up the *Select Data Source* dialog.

7. In the *Select Data Source* dialog click the *Add* button in the *Legend Entries (Series)* section.

8. Select the data for the *x* and *y* axes. You can also choose a cell containing a label for the variable.

9. Create the chart by clicking *OK* a couple of times.

You can edit the chart to make it clearer as required.

If you need to make another chart you can simply click once in the chart, then use the *Chart Tools > Design > Select Data* button to alter the data.

The *Chart Tools* menus give you many options; you can also right-click chart elements to access editing options for the selected element. You'll see more about charts in Part 2.

In the following exercise you can have a go at making a dot chart for some sample-based data which include a single predictor variable.

Have a Go: Make a dot chart for sample data

You'll need the *Pottery.xlsx* data for this exercise. There are two worksheets; the main data are in the *Data* sheet and the *Lookup* worksheet contains a simple lookup table to help make the dot chart.

↘ Go to the website for support material.

The data show the composition of pottery sherds from four sites. The first column, *Obs*, is simply an index number. The second column contains the site names. Columns C–G contain the concentration of different elements. The last column is an index based on a counter for each site.

1. Open the spreadsheet. You can see the main data in the *Data* worksheet. Go to the *Lookup* worksheet and view the simple lookup table. The first column contains the site names and the second is a number. These numbers will eventually be used for the *y*-axis in your dot chart.

2. Return to the *Data* worksheet. In cell I1 type a label for the *Site* number, Site.No will do.

3. Click in cell I2 and type a formula to convert the *Site* name from column B to an index number from the lookup table: =VLOOKUP(B2, Lookup!$A$2:$B$5, 2, FALSE). Note that you'll need the $ symbols. You can use the mouse to select the cells and then edit the formula afterwards, which is probably easier. Note also that you use FALSE to make sure

exact matching is used (so you do not need to have alphabetic order in the lookup table).

4. Once you are happy with the formula copy the cell I2 to the rest of the *Site.No* column. The data are already sorted in order of their *Site* name so it should be easy to see if you've got the formula correct.

5. Now you are going to make a scatter plot as a dot chart. Click in the Data worksheet; make sure you click away from any data, cell L7 is ideal.

6. Click the *Insert > Scatter* button. Choose the option with markers only (no lines), which should be the first icon (Figure 3.11).

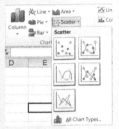

Figure 3.11 The *Insert > Scatter* button presents you with various scatter plot types.

7. You should now have a blank plot. Click once in the plot to make sure the *Chart Tools* menus are activated. Then click the *Design > Select Data* button.

8. Now click the *Add* button in the *Legend Entries (Series)* pane to bring up the *Edit Series* dialog box (Figure 3.12).

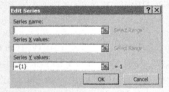

Figure 3.12 The *Edit Series* dialog box allows you to select data to populate a chart.

9. Click in the *Series name* box then click cell C1 to select *Al* as a label for the data series.

10. Click in the *Series X values* box, then select the values in the *Al* column (that is cells C2:C27).

11. Now click in the *Series Y values* box. Delete the entry that is already there, then select the cells containing the values for the *Site.No* (that is cells I2:I27). You'll see the chart forming in the background. Click *OK* to return to the *Select Data Source* dialog. Click *OK* again to close the dialog and return to the chart, newly populated. Your chart should resemble Figure 3.13.

You now have a dot chart representing the aluminium concentration in samples from the four sites. You can edit the chart to make it look nicer if you like. Since you are only going to use the chart to look for possible errors there is little point.

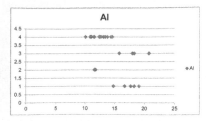

Figure 3.13 A basic dot chart showing aluminium concentration for samples from four sites.

You can see from your chart (Figure 3.13) that there is some variability but no single value appears to stand out as odd.

If you want to view another response variable you need to follow these steps:

1. Click once in the chart you wish to edit. This activates the *Chart Tools* menus.
2. Use the *Design > Select Data* button to open the *Select Data Source* dialog box.
3. Click the name of the series in the *Legend Entries (Series)* section to select that series.
4. Click the *Edit* button to open the *Edit Series* dialog box.
5. Delete the entry in the *Series* name box and select the label for the variable you want.
6. Now delete the entry in the *Series X values* box and then use the mouse to select the cells containing the variable you want.
7. Click *OK* to return to the previous dialog and *OK* again to close and return to the chart.

The scatter plot approach is useful but lacks flexibility; it is not easy to alter the variables displayed and if you add more data you need to select the data ranges all over again. If you have several predictor variables it can also be quite tricky to get the indexing correct. If you start from a pivot table however, you gain a great deal of flexibility, as you will see next.

Dot charts from pivot tables

Using a pivot table allows you to arrange and rearrange your data very easily. This gives you a lot of flexibility when it comes to making dot charts. Once you have a pivot table you can make a pivot chart from the data. Your chart cannot be a scatter plot; the nearest approximation is a line plot. This is what you want, as you can label the samples quite easily. However, you need to put in a bit of work to remove the lines, which can get in the way. Once you've made an appropriate chart you can save it as a template, which will be very useful for future dot charts and save you a lot of time.

It is possible to create a pivot chart without first making a pivot table; you simply use the *Insert >PivotTable> PivotChart* button. However, it is easier to see what is happening by going through the pivot table stage first. Once you've got the hang of it you'll be able to skip a stage (although it does not really save much time).

Arranging your data for a dot chart based on a pivot chart
Your final dot chart will end up arranged subtly differently from the one you made earlier (Figure 3.13); because of the limitations of the line chart you need to have the *x*-axis showing the categories, so the samples display vertically rather than horizontally (a line chart always shows the response variable on the *y*-axis).

Excel expects the first column of your data to be the categories. Subsequent columns are regarded as different *series*. This means that you need to arrange your pivot table so that the *predictor variable* is in the *Row Labels* section.

The rows of your pivot table will show the samples you want to chart. This means that you will have to make an index variable to allow the data to be arranged (see Section 2.1.1). You need to arrange your *index variable* in the *Column Labels* section when making the pivot table.

The *response variable* will form the *y*-axis and so this field will need to be in the *Values* section of the pivot table.

When arranged in this manner the line chart can display each of your data samples in a separate (labelled) column. Because the line chart treats each column as a series the points within each sample are shown with different markers, joined by lines. Having the lines in place does make the dot chart very hard to read so you need to remove them to leave the points. However, it is not so easy to alter the points. You can change each series' marker symbol but if you have a lot of replicates this becomes very tedious. It is not really worth the effort; you are going to use the chart to help you visualize the samples and help spot oddities (also called *outliers*), having different marker symbols does not alter the effectiveness of the dot chart.

It is best to make yourself a template chart that you can call up when you need it. You'll see how to do that now.

Making a dot chart template
You can make a template using any data, as long as it is laid out correctly. It is easier to make some data expressly for the purpose. All you need is a single column in a blank worksheet.

The process runs along the following lines:

1. Create a blank workbook in Excel.
2. Make a simple text label in cell A1.
3. Add some numerical values to the rest of the column. The easiest way is to enter the value 1 in cell A2 and 2 in cell A3. Then select both cells and use the drag handle to fill down the column. The sequence of values increases as you drag. Make 1000 or so; the exact number is not important.
4. Click once anywhere in (or adjacent to) the block of cells, then use the *Insert > Line* button to add a *Line with Markers*.
5. Right-click the data points and select *Format Data Series* from the popup menu.
6. From the *Format Data Series* dialog box select the *Line Color* section and use the *No line* option. Click *Close* to complete the operation.
7. You do not need the title so remove that; click it then press the delete key on the keyboard or use the *Chart Tools > Layout* dialog.

8. Delete the legend: click it and then use the delete key on the keyboard.
9. Click once on the chart to activate the *Chart Tools* menus. Then click the *Design > Save As Template* button.
10. Choose a sensible name for your chart template, then click the *Save* button to save it to disk. The folder location is selected automatically and all chart templates must be in this folder for Excel to be able to use them.

Once you have a chart template it can be used like any other chart. The *Insert* menu shows the charting buttons. They all have an *All Chart Types* option if you click on the triangle to expand the button (Figure 3.14).

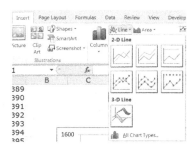

Figure 3.14 All charting buttons have an *All Chart Types* option, allowing you access to all the charts, including templates.

Your templates can be edited and modified like any other chart; you can even save the modified chart as a template.

Chart templates are stored in a special folder, which is not usually visible from *Windows Explorer*. You can access the folder most easily by using the *All Chart Types* option to bring up the *Insert Chart* dialog box (or *Change Chart Type* if a chart is open). You can now access the *Manage Templates* button, which opens the appropriate folder in an *Explorer* window. This allows you to copy templates into and from. In the following exercise you can have a go at inserting a chart template.

Have a Go: Add a chart template to Excel

You need the *NoLine.ctrx* file for this exercise. The file is a chart template, which you can use to make dot charts, that is line charts with no lines!

↘ Go to the website for support material.

1. With the *NoLine.ctrx* file downloaded, open a *Windows Explorer* window and navigate to the folder containing the file (use the *Start* button then *Computer*).
2. Now open Excel and you should be presented with a blank worksheet. Click the *Insert* tab on the ribbon and look at the *Charts* section.
3. Click the icon at the bottom right corner of the Charts section to open the Charts dialog box (Figure 3.15).

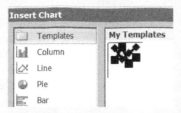

Figure 3.15 The bottom right of the Charts section has an icon that opens the Charts dialog box.

4. Now click the *Manage Templates* button in the *Charts* dialog box to open an *Explorer* window of the chart templates folder.
5. Click the *Taskbar* to bring up the Explorer window containing the *NoLine. ctrx* file.
6. Right-click the file and select *Copy* from the popup menu.
7. Click the *Charts* folder in the *Taskbar* to return to the chart templates folder.
8. Now right-click in an empty space and select *Paste* from the popup menu. The *NoLine.ctrx* file is copied into the folder.
9. You can now close the Explorer windows and return to Excel. The *Insert Chart* dialog box will still be open. Click the *Templates* section on the left and you should see your template listed and ready to use (Figure 3.16).

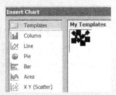

Figure 3.16 User chart templates are listed in the Templates section.

Note that the dialog box does not show template names. You can see the name of a template by hovering over the icon with the mouse. Now click the *Cancel* button and return to Excel.

Once you have your dot chart template you can use it to make diagnostic plots to help spot outliers in sample data.

You can make your dot charts by two main routes:

- Start from a pivot table then create the pivot chart.
- Make a pivot chart direct.

You'll see both approaches in the following sections, beginning with the indirect method.

Making a simple dot chart from a pivot table

You can use a pivot table as the basis for a pivot chart. The general way of going about things is as follows:

1. Make index variables as required to allow your pivot table to display the data in samples.
2. Create a pivot table. Usually you want the samples as the first column, which entails putting the *predictor* variable in the *Row Labels* section. The index variable needs to go in the *Column Labels* section. The *response* variable will be on the *y*-axis so this goes in the *Values* section.
3. Click on the pivot table to bring up the *PivotTable Tools* menus.
4. Click the *Options > PivotChart* button to start the process of chart building.
5. Select the appropriate chart type.
6. Edit the chart as required.

As you've already seen, you need to make a modified version of a line chart to use as a dot chart. It is easier to use a template that you've already built. In the following exercise you can have a go at making a dot chart using a chart template; you can either use the one you made yourself or the example one from the previous exercise.

Have a Go: Make a dot chart from a pivot table

You'll need the *Pottery.xlsx* data for this exercise. There are two worksheets. The main data are in the *Data* sheet. The *Lookup* worksheet contains a simple lookup table to help make an index based on the sites; you used this in a previous exercise but it is not required here.

↘ Go to the website for support material.

You'll also need the *NoLine.ctrx* chart template file. If you followed the earlier exercise this will be in your templates folder. Alternatively you can use your own template if you created one from the even earlier exercise (it will be in your template folder).

1. Open the spreadsheet and go to the *Data* worksheet. This contains the data with an index variable at the end (column H), which contains a site counter. This index will be used to make the columns of the pivot table.
2. Click once in the block of data and then click the *Insert >PivotTable* button.
3. The data should be selected automatically. Choose to place the pivot table in a new worksheet and then click *OK* to bring up the *PivotTable Field List* task pane.
4. Drag the *Site* field to the *Row Labels* section. You will see the site names listed in the pivot table that is forming.
5. Now drag the *Index* field to the *Column Labels* section and then drag the *Al* field to the *Values* section. The pivot table is now formed.
6. Click once in the pivot table to activate the *PivotTable Tools* menus. Now click *Options > PivotChart* to bring up the *Insert Chart* dialog box (Figure 3.17).

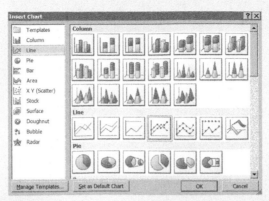

Figure 3.17 The *Insert Chart* dialog box allows you to select all available chart types, including user templates.

7. Ignore any template for now and select the *Line with Markers* option (fourth *Line* option). Click *OK* to insert the chart (Figure 3.18).

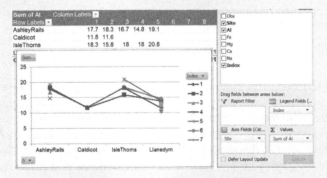

Figure 3.18 A pivot chart created from a pivot table as a line chart.

8. You don't want the lines in the chart. Notice also that it has additional buttons. These are called *Field Buttons,* which you can use to help *Sort* and *Filter* the data. You'll see how to use the buttons in detail another time. For now click on the chart then click the *PivotChart Tools > Design > Change Chart Type* button to bring up the *Change Chart Type* dialog box.

9. The *Change Chart Type* dialog box is identical to the *Insert Chart* dialog box, except for its title (Figure 3.17). Click the *Templates* section then click the *NoLine* template, or your own if you made one earlier. The icons are not labelled but you can see the name if you use the mouse to hover over the template icon. Click *OK* to apply the template.

10. You can use the *PivotChart Tools* menus to help you edit the chart as you like. Your final version should resemble Figure 3.19.

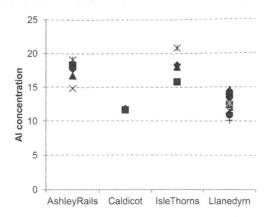

Figure 3.19 Dot chart to help explore data for outliers.

You don't really need axis titles as you are only using the data for validation, but it can be handy if you want to share your chart. You can see from Figure 3.19 that the data appear to be fine, nothing stands out. Try altering the variable that is displayed in the *Values* section of the *PivotTable Field List*. You'll have to reapply your template each time you alter the basic layout of the chart.

In the preceding exercise the data were fairly simple; there was only one predictor variable. When you have a more complicated situation you'll need to prepare your index variables carefully, so that you can display the data appropriately.

You will have noticed that the line charts display data using a range of colours. These colours can be somewhat distracting and it is generally helpful to be able to reduce the number of colours displayed. The *NoLine* template that you met earlier was made using an all-black theme. In the following section you'll see how to manage the colours.

Simplifying the colour scheme

You can manage the colours, fonts and themes used in Excel from the *Page Layout* menu on the ribbon. Excel has various colour schemes ready built in and you can experiment with these from the *Page Layout > Colors* button. Once you have a chart you can also apply a range of schemes using the *Chart Styles* section of the *Chart Tools > Design* menu.

When you make a chart template it takes on the colour theme that is in operation when you save it. This means that if you want a particular colour theme for your dot chart you have to save the template with that theme in place. If you modify the colour theme any existing charts are unaffected. An all-black theme is somewhat boring, but it is useful as it allows you to focus on the data and not be distracted by the colours.

It is easy to set the colour scheme so that all the elements are black:

1. Click on the *Page Layout* tab on the ribbon.
2. Click the *Colors* button in the *Theme* section; this shows the built-in themes.

3.   There is no all black theme so you'll have to make one. Click the *Create New Theme Colors* label at the bottom of the list. This opens a new dialog box.

4.   You want to set all the colours to black, except the two items labelled *Text/ Background – Light* (Figure 3.20); set them to white.

5.   Type a name for your theme in the *Name* box.

6.   Click *Save* to save your theme.

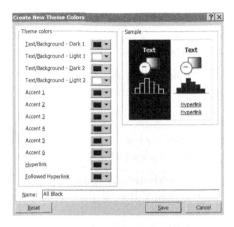

Figure 3.20 Setting an all-black theme is useful for dot charts with many replicates.

The custom colour themes you make are visible at the top of the list that appears when you click the *Colors* button, so they are available easily.

You can also change the font used via the *Page Layout > Fonts* button. This **does** affect any existing charts that you have in Excel, and they are altered immediately. Choosing a standard font such as Arial, Times New Roman, Verdana or Helvetica can be helpful if you want to save your charts as PDF files, since PDF editing programs do not recognize some of the Microsoft fonts.

A simple colour theme is especially useful if you have samples with a lot of replicates, as you'll see in the next section.

Complex dot charts created directly as pivot charts

The process involves a *PivotTable Field List* task pane as before, and a pivot table is created, but you build the chart rather than the table. It can be easier to work out how to arrange the fields using this route, as the *PivotTable Field List* task pane has subtly different labels, which can be easier to understand.

The process of creating a pivot chart directly is more or less as follows:

1.   Prepare your data with appropriate index variables.

2.   Click once in your main data then click the *Insert >PivotTable> PivotChart* button; the PivotTable button has a small "dropdown" triangle (Figure 3.21).

3.   The data should be selected automatically; choose to place the result in a new worksheet and click *OK*. The *PivotTable Field List* task pane should appear.

4.   You should see the *PivotChart Tools > Design* menu in the ribbon. Click the *Change Chart Type* button.

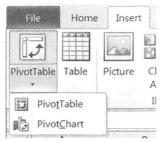

Figure 3.21 The *PivotTable* button can be expanded to show two options.

5. Select the style of chart you need; this will probably be in the *Templates* section.

6. The *PivotTable Field List* task pane shows slightly different headings from when you are inserting a pivot table. Drag the fields into the appropriate sections. You'll want the *predictor* variable(s) in the *Axis Fields (Categories)* section and the index variable (or whatever forms the replicates) in the *Legend Fields (Series)* section. The *response* variable goes in the *Values* section.

The chart forms as you drag the fields around the *PivotTable Field List* task pane, so you can experiment easily. This also means it is easy to explore your data because you can alter the display quickly.

In the following exercise you can have a go at making a pivot chart directly.

Have a Go: Make a pivot chart directly

You'll need the *warpbreaks.xlsx* data file for this exercise. The file contains two worksheets; *Recording-layout* contains the basic data and *Recording-layout-indexed* includes extra index columns. The data represent the number of breaks in looms of wool under various tensions. There are two types of wool and three tension settings.

↘ Go to the website for support material.

1. Open the *warpbreaks.xlsx* spreadsheet and go to the *Recording-layout-indexed* worksheet.

2. Click once anywhere in the block of data then click the *Insert >PivotTable> PivotChart* button (click on the triangle on the *PivotTable* button). The data should be selected automatically, so choose to place the result in a new worksheet and click *OK*.

3. The *PivotTable Field List* task pane is open; note that the labels on the sections are slightly different to when you are making a pivot table. The default chart is usually a bar chart so click the *PivotChart Tools > Design > Change Chart Type* button to open the *Change Chart Type* dialog box (identical to Figure 3.17).

4. Click the *Templates* section and then select the *Noline* template (hover over the icons with the mouse to display the template name). Click *OK* to apply the template; you can now build the chart.

5. Start the chart-building by dragging the response variable (*Breaks*) from the list at the top of the *PivotTable Field List* task pane to the *Values* section at the bottom.
6. Drag the predictor variables to the *Axis Fields (Categories)* section. You want *Wool* and *Tension*. The order is not important; you can change it later, but start with *Wool* above *Tension*.
7. Drag the *Sobs* index variable to the *Legend Fields (Series)* section. The chart is now complete (Figure 3.22).

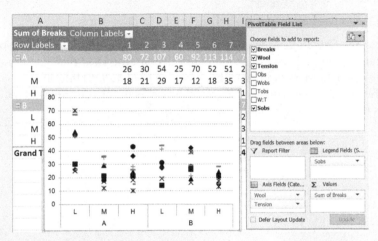

Figure 3.22 A dot chart made directly as a pivot chart.

You can use the *PivotChart Tools* menus to help you format the chart. You can also use right-click on many of the chart elements. Keep the file open for the moment, as you'll look at the pivot chart in more detail shortly.

Once you have created a pivot chart you can edit it like any other chart. However, there are also a few things that you can do that you cannot undertake with a regular chart. You can easily rearrange the chart using the *PivotTable Field List* task pane, although you'll have to reapply your dot chart template. You can also carry out *Filter* and *Sort* operations; these help make your pivot chart a powerful and useful diagnostic tool, as you'll see next.

Using a dot chart as a diagnostic tool

A pivot chart usually has some additional buttons, these relate to the fields from the *PivotTable Field List* task pane (see Figure 3.18). These buttons allow you to edit the fields and alter settings with a right-click. If the button contains a triangle you can click it to bring up the *Filter* dialog. This can be useful as the *PivotTable Field List* task pane generally gets in the way; you can turn it off to give more room on the screen (use the button at the top right of the pane). To turn it on again you can click once in the pivot table then use the *PivotTable Tools > Options > Field List* button. You can also access the *Field List* and *Field Buttons* from the *PivotChart Tools > Analyze* menu (Figure 3.23).

Figure 3.23 The *Analyze* menu allows access to useful tools for pivot charts.

You can use the *Field Buttons* on a pivot chart to apply *Sort* and *Filter* operations. If you make a chart template it will most likely not display the *Field Buttons,* so you need to be able to switch them on (and off). The *Analyze > Field Buttons* button is one method to manage the *Field Buttons;* you can also click the fields in the four areas of the *PivotTable Fields* task pane to display a popup menu (Figure 3.24).

Figure 3.24 Clicking a field in the *Pivot Table Field List* brings up a dialog, allowing access to the field buttons.

Once you have the *Field Buttons* displayed on the chart you can close the *PivotTable Fields List* task pane, allowing you more space on screen to explore the chart.

If you click a *Field Button* you get a dialog allowing you to apply some *Sort* and *Filter* operations (Figure 3.25).

However, the *Sort* options are limited; you cannot apply a custom list for example. If you click on a *Field List* button one option reads *More Sort Options.* This gives you some extra control, including a cryptic *Manual* sort (this is the default). To apply a manual sort, you need to rearrange the underlying pivot table. Right-click in one of the row label items to bring up a new dialog; one option is *Move* (Figure 3.26).

You can shift the item using this right-click approach and your pivot chart will update to reflect the new arrangement.

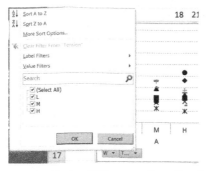

Figure 3.25 *Field Buttons* allow access to *Sort* and *Filter* operations directly from a pivot chart.

Figure 3.26 Right-clicking a heading in a pivot table allows you to move the item, thus applying a custom sort.

You can use the filters to focus on certain samples and look at the spread of the data. Remember that you are looking for data points that look odd. This does not mean that the data are incorrect, but it does allow you to screen for potential errors.

In the following exercise you can have a go at using the *Field Buttons* for yourself. You should already have an open spreadsheet containing a dot chart as a pivot chart. If you haven't got the file open you can re-run the preceding exercise to make the pivot chart.

Have A Go: Use field buttons and sorting on a pivot chart

You'll need the results of the preceding exercise for this. The data file is *warp-breaks.xlsx*, which shows the effects on breakage of tension for two sorts of wool yarn. You should have created a pivot chart from the data (as a dot chart) using a template.

1.   Click once on the chart to activate the *Pivot Chart Tools* menus. Then click the *Analyze* tab then the triangle on the *Field Buttons* button (see Figure 3.23).

2. From the popup menu click the *Hide All* option to uncheck it. Now the other options on the *Field Buttons* menu will turn the field buttons on or off. Click each option to enable all the field buttons.
3. Click the *Tension Field Button,* this will be the one on the right (the name may be not visible if the button is small). A dialog appears. Apply a *Filter* to the data: display just the "L" tension. Click *OK* to apply the *Filter* and return to the chart.
4. Look at the *PivotTable Field List* task pane. You'll see that the *Tension* field has a funnel icon, indicating that a *Filter* is in operation (Figure 3.27).

Figure 3.27 If a *Filter* is in operation a funnel icon is visible in the *Field List*. The *Filter* can be accessed by clicking on the field.

5. Click to the right of the funnel by the *Tension* field item. From the popup menu select the *Clear Filter From "Tension"* option, to remove the *Filter*.
6. Now click the *Tension Field Button* again and apply a *Sort* from Z to A. Notice that the icon in the pivot table itself changes to show you that *Filter* or *Sort* operations are in force.
7. Click an *M* in the pivot table; it does not matter whether you click the one in the *A* or the *B* wool sections. You may have to move the chart to get to the pivot table.
8. Now right-click to bring up a menu, and hover over the *Move* option. A further popup menu appears showing various ways you can shift the item; select *Move "M" Up*. The pivot table and chart are updated immediately.

Try out other *Sort* and *Filter* options. You might also rearrange the data by right-clicking in a column heading.

Note: Slicer tool
You can use a slicer as an interactive filter for data in pivot tables. You can insert a slicer from the *Insert* or *PivotTable Tools > Options* menus. You'll see the *Slicer* tool in more detail in Chapter 6.

The *Field Buttons* are helpful in allowing you to explore your dot chart, especially when you have large datasets. Remember that you are looking for possible errors; just because a datum looks a bit out of place does not mean that it is a mistake, but you should look again more carefully and make sure. The dot charts you create are diagnostic tools to help you validate your data. They are not really suitable as a way to summarize your data; you'll explore these methods in Part 2.

The processes you've seen so far have pertained to data that is split "by sample", that is you are comparing samples and looking for differences. When you are looking for links between variables, that is correlations and regression, you need a different approach, which you'll see in the next section.

3.3.2   Errors in correlation data

If you are looking for links between variables; that is correlation or regression, your data are most often numeric variables although you may also have other non-numeric factors.

The usual way of presenting correlations visually is to use scatter plots. You'll see more about the presentation of results in Part 2 but the scatter plot is also a useful validation tool.

The processes of correlation and regression enable you to see how good the fit between two variables is. Each datum has an influence on the fit. If a point is way off, it alters the regression. The influence of outliers depends on various factors, such as how far off the average they are and how large the dataset is. In general the smaller the dataset (that is the fewer replicates you have) the larger the influence of rogue data. However, don't be too quick to throw away data because they look odd, sometimes these oddities are the interesting bits!

Scatter plots allow you to see one variable plotted against another. The *y*-axis is usually the *response* variable and the *x*-axis is usually the *predictor* variable(s) but other combinations can be useful. Making scatter plots allows you to explore your data and check for outliers; it also allows you to see potential patterns in the data; this exploration is sometimes called *data mining*. In the following sections you'll see how to explore your data using scatter plots. The focus is on data validation but the skills you learn here will also be useful for general data visualization (see Part 2).

Scatter plots from columns

You can make scatter plots using the *Scatter* button, which is in the *Charts* section of the *Insert* menu on the ribbon. The options appear when you click the button. The *Scatter with only Markers* option is what you'll use most of the time (Figure 3.28).

You need data to make your scatter plot and there are several ways to go about making your charts:

- Select the data you want using the mouse.
- Click once in the data and let Excel select the data.
- Select no data, make a blank plot and choose the data afterwards.

Figure 3.28 The *Scatter* button allows several sorts of scatter plot; the option for markers (points) only is the most commonly used.

Using the mouse to select data is the main way that most people set about creating any chart. If your data are arranged appropriately this is fine but often your data are not arranged how Excel expects. If you have a large dataset it can be tedious to select many rows, although there are shortcuts.

Allowing Excel to choose the data for you is usually the worst option. The program has excellent charting power but the layout and starting point for charts is woefully inadequate. It can be quite difficult to get your data into the right order for Excel to understand. The latest version (Excel 2013) has improved this aspect but chart-building is some way behind pivot table creation for example.

As a general rule it is best to start from a blank chart and populate it yourself. This gives you good control over the data and allows you to create the chart you want. The drawback is that it usually takes a bit longer than the other options. If you are making a chart for presentation it is generally worth the extra effort.

For simple data validation you probably want a faster approach. In the following sections you'll see various ways to make scatter plots to use for data mining and validation.

Selecting column data

If your data is arranged in columns, with each column being a separate variable, it is easy to select variables using the mouse. It's usual to have your *response* variable on the *y*-axis and the *predictors* on the *x*-axis, so immediately you have a problem. Excel expects your data to have an *x*-variable and one or several *y*-variables, with the *x*-variable first. You can get over this issue only by building the chart and specifying the data afterwards.

Since the plots will be used for validation purposes you can accept this limitation, which does not affect your ability to spot outliers. You need to select your data using the mouse, and then use the *Scatter* button to build a chart. There is no need to click and drag down multiple rows to select your data; you can simply select an entire column by clicking in the header (the column letter). The general process goes as follows:

1. Click the column header of a variable you want to plot; the column on the furthest left is used as the *x*-axis regardless of what other columns are selected (or the order you select them).
2. Use Ctrl+Click to select subsequent columns. If your columns are all adjacent you can click and drag to extend the selection over several columns. You can also use Shift+Click to select columns between two points (that is, click one column, then Shift+Click another to select all columns between the two).

3. Click the *Insert > Scatter* button then select the *Scatter with only Markers* option. The chart is created immediately.
4. Edit your chart to make it more readable if necessary using the *Chart Tools* menus.

This process allows you to make scatter plots quickly and easily. If you want to make a new plot you can simply click the columns you require and make one. In the following exercise you can have a go at making some scatter plots.

Have a Go: Make scatter plots for data validation

You'll need the spreadsheet *NYenvironmental.xlsx* for this exercise. The data shows Ozone measurements from New York in 1973 (May to September). There are also columns for solar radiation, temperature and wind speed.

↘ Go to the website for support material.

1. Open the *NYenvironmental.xlsx* spreadsheet; there is only one worksheet, *Data*. The first column contains a reference number containing the observation. Use the mouse to select the *Ozone* column (click in the header bar). This will form the *x*-axis although it is really the *response* variable.
2. Now Ctrl+Click the header bar of the *Temp* column to select that. You should now have *Ozone* and *Temp* columns highlighted.
3. Click the *Insert > Scatter* button and select a chart with markers only. The chart is created immediately. You'll see that the chart has a title and a legend (Figure 3.29).

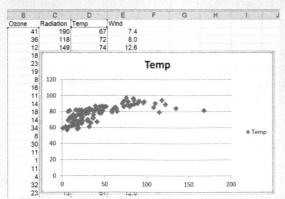

Figure 3.29 A scatter plot with a single variable on each axis displays a main title and legend by default.

4. Delete the chart; click once somewhere near the top right of the chart then use the *Delete* key on the keyboard.
5. Click the *Ozone* column header and drag the mouse to the right to select columns B to E (that is, all of the data).

6. Insert a new scatter plot with the *Insert > Scatter* button. This time you'll see that there is no title but the legend allows you to differentiate the data series (Figure 3.30).

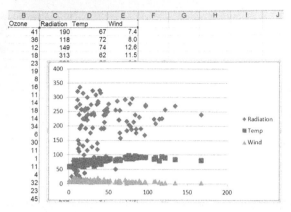

Figure 3.30 A scatter plot with multiple *y*-axis variables displays a legend by default.

Try different selections and see how easy it is to select the columns you want to plot.

It does not matter in which order you select variable columns; the leftmost column is always treated as the *x*-axis variable. You could move the column to the rightmost end but then you still have the problem that Excel will only accept a single column as an *x*-variable.

You can force Excel to treat columns as you want, by building scatter plots (that is specifying the data you want included yourself), as you'll see shortly. For the purposes of checking for outliers and generally looking for patterns in the data it perhaps does not matter so much if *x* and *y*-axes are transposed.

Comparing charts

It can be useful to be able to see multiple plots at once, so that you can compare them more easily. You can achieve this by dragging and resizing your charts with the mouse. This is tedious and if you add more charts the rest need resizing. The answer is to move each chart to a new worksheet and then arrange them on screen.

When you have a chart selected you'll see the *Chart Tools* menus. The *Design* menu contains a *Move Chart* button, which allows you to move the chart to a new worksheet. Once you have several charts you can use the *Window* section of the *View* menu to arrange your charts. The general way to proceed is as follows:

1. Make a chart. Click once on the chart to activate the *Chart Tools* menus, then click the *Design > Move Chart* button (Figure 3.31).

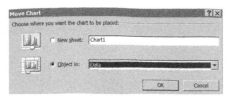

Figure 3.31 The *Move Chart* dialog allows you to move a chart to a new worksheet.

2. Select the *New sheet* option. You can type a new name for the worksheet if you like. Click *OK* to move the chart.
3. Return to the main data and make and move your next chart (steps 1–2).
4. Continue until you have as many charts as you need, each one in its own worksheet.
5. Click the worksheet tab that contains the first chart you want to look at. Now click the *View* menu and then the *New Window* button.
6. The *Windows* taskbar shows a new active window, and you see a second Excel window icon, with the name of the spreadsheet and a number (Figure 3.32).

Figure 3.32 Adding multiple windows creates additional icons in the *Windows* taskbar.

7. Click the *New Window* button as many times as you need for the number of charts you want to compare.
8. At this point each of the *taskbar* windows shows the same chart. Click the first taskbar icon and then select the worksheet tab that you want to display.
9. Click each *taskbar* icon and select the corresponding chart. You should end up with a different chart in each *taskbar* window.
10. Now from the *View* menu click the *Arrange All* button; this presents you with several options (Figure 3.33).

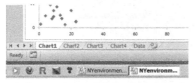

Figure 3.33 The *Arrange Windows* button (via *View > Arrange All*) allows you to display multiple windows in various ways.

11. You want to have the windows *Tiled*. Select the *Windows of active workbook* check box, and then click *OK* to arrange the selected charts on screen (Figure 3.34).

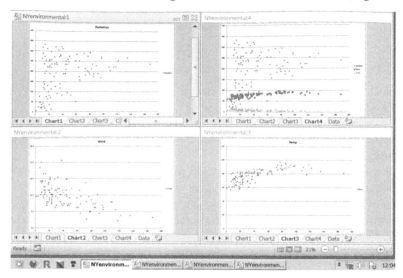

Figure 3.34 Multiple charts can be arranged on screen for comparison if they are in separate Excel windows.

The separate windows act like regular Excel windows; you can minimize, maximize or close them using the icons at the top-right of each window. Remember that you should click once in the header of a window to make it the active window before you try to undertake any tasks.

You can select different worksheets from each of the open windows, so if you have four charts on screen but others not displayed, you can choose (from the worksheet tabs) which to present. The more charts you have on screen at any one time the smaller they'll be.

If you are looking for outliers you'll probably want to focus on one chart with one pair of variables at a time. Being able to switch the displayed variables would then be a useful ability. You can choose a more dynamic variable selection, as you'll see next.

Dynamic selection
If you are looking for outliers you will most likely display a single variable on each axis. There are a couple of simple ways you can switch the variables that you display:

- Use the data selection outlines.
- Hide and unhide data columns.

If you make a chart then click on it you'll see that the data used in the chart is subtly highlighted. The various elements are outlined by a thin coloured border (Figure 3.35).

If you hover over these borders with the mouse you can see the cursor change: you can drag some of the data elements to alter the range or move a selection. In practice you can really only alter the values displayed on the y-axis. There are two main ways:

	B	C	
	Ozone	Radiation	Temp
	41	190	
	36	118	
	12	149	
	18	313	
	23		
	19		
	8		

Figure 3.35 When a chart is selected the data elements are outlined.

- Hover over a top corner until the cursor alters to a double-headed arrow. Then click and drag to expand the selection to adjacent columns. This adds more variables to the chart.
- Hover over an edge until the cursor alters to a four-way arrow. Then click and drag to move the selection to a new column. This changes the variable displayed on the *y*-axis.

You cannot alter the leftmost column, the *x*-axis, even though the cursor alters, except to reduce the number of rows.

> **Note: Dynamic selection and Excel options**
> The selection borders do not appear if *Drag and Drop* selection is disabled. You can enable drag and drop editing from the *File* menu on the ribbon. Select *Options > Advanced* and tick the box beside the heading that reads *Enable fill handle and cell drag and drop*. Now when you click a chart the selection borders should be visible.

The other way to switch variables quickly is to select all the variables you want to explore and make your chart. Then you *Hide* the columns you do not want to see and *Unhide* the ones that you do want.

You can access the *Hide* and *Unhide* buttons most easily using a right-click. To *Hide* columns you select the columns you want to hide then right-click and choose *Hide* (Figure 3.36). You can also use the *Home > Format* button.

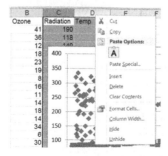

Figure 3.36 Columns can be hidden to remove them from a chart using the *Hide* option.

Once columns are hidden you'll notice that the column headings show a gap; if you *Hide* columns C and D you'll see the column headings jump from B to E. To *Unhide* columns you select the columns either side of the gap, then right-click (or use the *Home* > *Format* button). The *Unhide* option does not work if the selection does not contain any hidden columns.

> **Tip: Charts and hidden columns**
> If you make a chart then hide some columns the chart alters in size. To maintain a clear display you can place the original chart in a new worksheet. Make a new window and arrange it so that the data and chart are arranged horizontally. You can resize the windows using the mouse.

Dragging selection borders and hiding columns are useful ways to manage your plots and allow you some freedom in exploring your data. You can achieve a higher degree of control by building a plot from scratch, which you'll see next.

Building scatter plots

If you start with an empty chart you can populate it with data as you like. This gives you more control over which variable is placed on which axis. You must start by ensuring that the current insertion point (that is, the cell that is active) is not in, or adjacent to, any block of data. This is because Excel assumes that any data nearby is what you want to use. In some cases this can be useful (as in making pivot tables) but in this case it is not a desirable effect. Once you have a blank chart you can add data as you like. The general running order is:

1. Click on a cell away from any data (even on a separate worksheet).
2. Use the *Insert* menu to make the chart of your choice.
3. Click on the chart to activate the *Chart Tools* menus.
4. From the *Design* menu use the *Select Data* button to bring up the *Select Data Source* dialog box.
5. Use the *Add* button to add data series. You can select the x and y variables as well as a name for the series (the name relates to the x-axis).

Once you've added data series to your chart you can use the *Edit* button in the *Select Data Source* dialog to alter the data series or use the *Remove* button to delete a series from the chart.

> **Note: Hidden columns and built charts**
> If you make a chart using the building process (that is, using the *Select Data* button) you can insert one or more data series. If you subsequently hide a column the variable is still displayed on the chart (although not displayed in the worksheet).

When you build a chart in this fashion you have to select the data variable by variable. If you are error checking you will probably plot only one data series at a time, that is, one response variable (*y*-axis) and one predictor variable (*x*-axis).

If you need to check other variables you'll have to edit the data series. There are three ways to do this:

- Use the *Select Data* button and the mouse to edit or select new variables.
- Use the dynamic selection method by dragging the data selection borders.
- Use named ranges, which allows you to select data without using the mouse.

The first option is usually the quickest only if your data contains few rows. If you are starting from scratch using named ranges can be very useful, as you'll see later. However, if you've already created a chart then you might as well use the dynamic selection method.

Selection borders

If you've built a blank chart and then populated it with data you can't see the data selection borders by simply clicking on the chart, as you did earlier (Figure 3.35). To view the data selection borders you need to:

1. Have the chart in the same worksheet as the data.
2. Click once on a data point in the chart.

Once you've activated the data selection borders you can use the mouse to move the variables represented on the chart. You can move the *x* and *y* variables as well as the name for the series (presuming that you clicked a cell containing the variable name).

Usually you'll have one response variable and several predictors. Using the selection borders allows you to redraw your chart quickly and visualize the relationship between the response and each predictor in turn; looking for errant data points that might be errors.

Tip: Highlighting a single data point

If you click on a chart it is made active and the *Chart Tools* menus become available. If you click on a data point you'll activate the data series and the selection borders will be visible (if the chart is in the same worksheet as the data).

If you click a data point a second time you select that datum only. This allows you to alter the colour, for example, thus highlighting a possible outlier in your dataset. If you then alter the response variable using the selection border the point will keep the same format, allowing you to follow it through different predictor variables.

Using selection borders works well if you have already built a chart; however you would have to select the data originally using the mouse. If you have a lot of data rows using the mouse can be tedious. Also, your chart has to be in the same worksheet as the data for the selection borders to be visible, which can be inconvenient. An alternative approach is to use named ranges right at the outset. You'll see this approach next.

Named ranges

It would be nice to be able to select the entire column, as you did before, but this does not work for chart-building. When you select the entire column you include the header but in the *Select Data Source* dialog you need to define the name separately from the data. What you need to be able to do is assign a name to the range of cells that contains the data of a variable but not the heading. You can then use the name in the *Select Data Source* dialog instead of highlighting the cells with the mouse.

Excel allows you to assign names to ranges of cells. You can find the tools for assigning names in the *Defined Names* section of the *Formulas* menu (Figure 3.37).

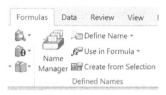

Figure 3.37 The *Formulas* menu contains a *Defined Names* section, allowing you to manage named ranges.

One way to assign a name is to highlight a range of cells with the mouse then use the *Formulas > Define Name* button. However, the idea is to avoid having to use the mouse so you need a quick way to assign names to all the variables in your dataset. You need to use the *Create from Selection* button, which you operate thus:

1. Make sure your data are in scientific recording format, with each column (variable) having a unique name. It is useful if the names are short but meaningful.
2. Click once anywhere in the block of data. You must be in the block and not adjacent, which was okay for pivot tables but not for this.
3. Press Ctrl+A on the keyboard to select all the data.
4. Click the *Formulas > Create from Selection* button to bring up a new dialog box (Figure 3.38).
5. Use the mouse to select the source of the name labels; this is the first option labelled *Top row* (other items may be selected as shown in Figure 3.38). Click *OK* to complete the operation.

Figure 3.38 Creating named ranges automatically from column headings.

Once you've created the named ranges from your dataset you can view the results using the *Formulas > Name Manager* button, which opens the *Name Manager* dialog box (Figure 3.39).

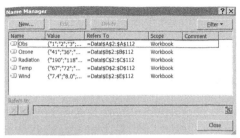

Figure 3.39 The *Name Manager* dialog box allows you to manage the named ranges in your workbook.

You can use the *Name Manager* dialog to alter the name of the items, to alter the cell range or to add comments. You can also delete named range items. If you add more data later you simply repeat the steps above and replace the existing definitions.

Once you have the named ranges defined you can use the names instead of cell references in formulae, and of course in creating charts. Note however, that you cannot just type the name into the *Edit Series* dialog box. If you type a simple name Excel regards it as plain text; what you must do is append the name of the worksheet with an exclamation mark. So if you have a worksheet called *Data* and a variable called *Ozone* you type `Data!Ozone`.

In summary the process of using named ranges in building a chart is as follows:

1. Create your named ranges using the *Formulas > Create from Selection* button.
2. Make a blank chart from the *Insert* menu. Make sure you click on an empty cell that is not adjacent to any data. You can use a separate worksheet to make it easier to access an empty cell; arranging windows so that you can see the data and the blank worksheet is also helpful (use the *View* menu).
3. Click once in your empty chart to bring up the *Chart Tools* menus.
4. Click the *Design > Select Data* button to open the *Select Data Source* dialog box.
5. Click *Add* to open the *Edit Series* dialog box.
6. Type a name in the *Series name* box. This is for your convenience and will form the title and legend. You could select a cell but it is easier to type something.
7. In the *Series X values* box type the name of the worksheet that contains the data, append with an exclamation mark and then append with the name of the predictor variable. The name should match the named range you defined (Figure 3.40).
8. Repeat step 7 but use the *Series Y values* box and the response variable (Figure 3.40).
9. Click *OK* to return to the *Select Data Source* dialog and *OK* again to return to the worksheet and view your completed chart, which you can edit as you need.

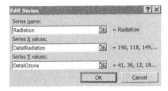

Figure 3.40 Named ranges must have the worksheet appended with an exclamation mark to be used in naming cell ranges for charts.

If you return to edit a chart you'll see that the name of the worksheet has been replaced with the name of the workbook. Your defined names are effective across all the worksheets of your workbook so Excel alters the names internally. However, the names of your worksheets are usually shorter than the filename so type whatever is shortest!

In the following exercise you can have a go at building a scatter plot using named ranges.

Have a Go: Build and edit a scatter plot using named ranges

You'll need the spreadsheet *NYenvironmental.xlsx* for this exercise. The data shows Ozone measurements from New York in 1973 (May to September). There are also columns for solar radiation, temperature and wind speed.

↘ Go to the website for support material.

1. Open the *NYenvironmental.xlsx* spreadsheet; there is only one worksheet, *Data*. The first column contains a simple reference number containing the observation. The other columns are *Ozone* (the response variable) and others representing three predictor variables, *Radiation, Temp* and *Wind*.
2. Click once somewhere in the block of data, then select everything using Ctrl+A on the keyboard. You should see the data highlighted.
3. Now click the *Formulas > Create from Selection* button to open a dialog box. Make sure the box labelled *Top row* is ticked and that no others are. This will ensure that the top row is taken as labels. Now click *OK* to apply the selection.
4. Click on the *Name Manager* button to see the named range items. Click *Close* when you have seen the items.
5. Now click once somewhere outside of the data block, and not adjacent to it; near the top of column G would be ideal. Then click the *Insert > Scatter* button and select a scatter plot with only markers. A blank chart is created immediately.
6. Click once on the chart then use the *Design > Select Data* button to open the *Select Data Source* dialog box.
7. Click the *Add* button to open the *Edit Series* dialog box. In the *Series name* box type `Radiation`.
8. Now click in the *Series X values* box and type `Data!Radiation`.
9. Click in the *Series Y values* box and type `Data!Ozone`. Click *OK* to return to the *Select Data Source* dialog box and *OK* again to return to the worksheet. Your chart should be visible.

10. Since you are only using the chart to look for potential errors you do not need to spend too much time editing the chart. However, you do not need the legend since you are only plotting one predictor variable. Click on the legend and use the Delete key on the keyboard to remove it (you can also right-click and select *Delete*). Your chart should resemble Figure 3.41.

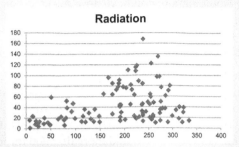

Figure 3.41 Scatter plot built using named ranges to check for outliers.

11. Click once on the chart then click the *Design > Select Data* button. Click the name of the data series and then the *Edit* box from the *Select Data Source* dialog box.

12. Alter the *Series name* to Wind. Alter the *Series X values* to Data!Wind. Leave the *Series Y values* as they are (showing the *Ozone* variable). Click *OK* to return to the previous dialog and *OK* again to return to the worksheet. The chart should now show the new data.

13. Now click on the chart and then click once on any data point. You should see the selection borders appear in the worksheet around the variables used for the graph. Hover with the mouse near to a border of the *Wind* column (the *x* values) until the cursor alters to a four-headed arrow. Click and drag the selection borders to the *Temp* column. When you release the mouse button the graph updates but the title is incorrect.

14. Because you typed text into the *Edit Series* dialog (the *Series name* box) no cells are highlighted. However, the formula bar displays the current data range (Figure 3.42).

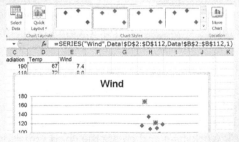

Figure 3.42 Data ranges used in charts appear in the formula bar. Here the name of the series does not match the variable used and needs to be edited.

15. Click in the formula bar and alter the *Wind* item to read Temp. Once you press Enter the chart title updates. You can also alter the title by clicking it on the chart and typing the new name.
16. Click in the corner of the chart so that no points are active. Now click once on the data point furthest up the *y*-axis; you have selected the entire data series. Click the point again; you have now selected that single point.
17. Double click the point and make the point a different colour: select *Marker Fill* then *Solid Fill* then the colour (red will do nicely). Click *Close* to apply the formatting to that point.

You've now applied a different formatting to a potential outlier. If you hover over the point you'll see the values associated with the point. Try altering the response variable and see how the same point shows up. Although this datum is slightly apart from the others you have no reason to suppose that it is an error. If you had collected the data yourself or had more detailed information you might be able to find a reason for the exceptionally high ozone level. You might make a note of the reading so that you can see the effect removing it might have on any statistical analyses later on.

In the preceding exercise you made a chart using named ranges. These allow you to build charts fairly rapidly, without having to use the mouse to highlight lots of data. You can alter the chart using other named ranges or use the selection outline. However, the selection outline only works if your chart is in the same worksheet as the data. If you have a lot of data variables it is probably easier to have your chart in a window of its own, making named ranges the easiest method to use to alter the chart.

Having your charts in a separate worksheet is generally useful. It is better to keep your data separate from any analysis. You can simply move the chart to a new worksheet (or create it in one) and maintain separate windows so that you can view the chart and data at the same time. One occasion when this is especially useful is when you apply filters to your chart data, as you'll see next.

Filtering data in scatter plots
You can apply *Filter* operations to data that are linked to charts. If you do so then the chart is updated immediately to reflect the data on view. Since the *Filter* operation hides certain data any chart alters in size or disappears altogether unless it is in a separate worksheet.

If you want to apply a filter to a chart then the general process is as follows:

1. Make a new worksheet to accommodate the chart.
2. Make a new window from the *View* menu and arrange the windows so you can see the data and the blank worksheet (use the *Arrange All* button).
3. Click in the blank worksheet and create the chart you require.
4. Click the *Data* tab and use the *Data > Filter* button to turn filters on.

5. Click the triangle icons in the column headings to bring up the Filter dialog box. Your chart updates as soon as the filter is applied.

The *Number Filters* menu in the Filter dialog is especially helpful. There are several useful options that you can use in error checking:

*   *Above Average* – this shows data that lie above the mean, so you see the top end.
*   *Below Average* – this shows the bottom end.
*   *Top 10* – this shows you the top (or bottom) number (or percentage) of items. Although labelled *Top 10* you can alter the number (Figure 3.43), so you could view the top 20 percent or bottom 5 ranked for example.

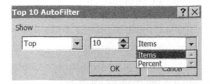

Figure 3.43 The *Top 10* filter allows you to filter the ends by rank or percentage.

These filters can be useful in error checking, particularly if you have a lot of data points. You are generally interested in the more outlying items when it comes to errors. Of course other intermediate values could be errors but you have no way of spotting them. The point of data validation is to make sure that you have the best data you can get. If something looks strange it could be because a decimal point was in the wrong place or because the equipment had a glitch. On the other hand it could simply be that your data are variable!

3.4 EXERCISES

1. Which of the following tools are useful when dealing with typographical errors?

    A) Sort tools.
    B) Pivot tables.
    C) Filter tools.
    D) Chart tools.
    E) Data validation tools.

2. You have to use a pivot chart when checking for numerical errors. TRUE or FALSE?
3. A ____ chart is useful for spotting numerical oddities in differences data. You can make a template starting from a ____ plot.
4. Which of the following options are NOT correct with regard to the data validation tools?

    A) Users can select from a list of options.
    B) You can restrict entry to whole numbers.
    C) You can allow any value with a warning.
    D) You can display a guide message at input.
    E) You can set error messages to display on an already completed dataset.

5. Look at the *Error Checking.xlsx* file. The *Species* column contains names, some of which have additional spaces at the end. How can you use a *Filter* to display a list of all records containing *Species* with these trailing spaces?

↘ Go to the website for support material.

The answers to these exercises can be found in Appendix 1.

3.5 SUMMARY

Topic	Key points
Types of data error	Data can be in error for two main reasons: a typographical error (for example, a spelling mistake) or a numerical error (for example, the wrong value). Spelling mistakes are easier to deal with.
	A text item can have additional spaces at the end and appear to be correct.
	Numerical oddities are harder to spot. Just because a value is odd does not mean it is necessarily incorrect.
Sorting tools	Sorting tools allow you to arrange items in order; misspelled words are thus more easily spotted.
Pivot tables	Pivot tables can help you spot typographical errors as items spelt slightly differently appear as separate headings. If two items appear identical it is likely that one has additional spaces at the end.
	Find and Replace tools can be used in the main dataset to search (and replace) items you spot as errors in the pivot table.
Filter tools	Filter tools can show similar entries; you can then apply the filter and correct the entries in place.
	The *Advanced* filter can be used to find additional spaces at the end of all entries. You can use the EXACT and RIGHT functions to help apply the filter.
Data validation	Excel has data validation tools that allow you to restrict entries. You can specify ranges for numerical values or supply a list of valid options, for example. You can set up messages that will display to guide the user. For text entries you need to produce a lookup table containing the allowable items.
	Data validation can be used after data entry to highlight entries that lie outside of the criteria you set.
Graphical methods	Graphs are a good way to scan for errant data. Differences data require dot charts, which you can construct from a line plot as a starting point. Correlation and regression data need a scatter plot.
	Data that look odd may not be real errors but may simply represent variable data. Check carefully and only remove or alter data when you are sure that a datum is a real error.

Topic	Key points
Sample data	Dot charts are helpful to show the data sample by sample. Using a pivot table makes this process flexible since you can rearrange the data and apply filters as you need. Pivot charts contain *Field Buttons*, which allow you to filter data easily.
	You can make a dot chart template by creating a line plot and removing the line. Then save the chart as a template, which you can use on data in the future.
Reorder plot labels	If you right-click on a heading in a pivot table you can alter its position, thus creating a custom sort order.
Correlation data	Scatter plots are most suitable for correlation and regression data. You must make the scatter plots direct (that is, not as pivot charts). You can alter the selected data easily using the selection borders, which appear in the worksheet when you click on the chart. Filters allow you to view subgroups in your data.
Hiding and filtering	You can hide columns you don't want to appear in a chart. This allows you to build a chart with several variables. You can select which variables appear by selectively hiding columns.
	If you apply a filter the chart updates to reflect the filter.
Multiple windows	You can make extra windows using the *View > New Window* button. You can arrange the windows on screen, allowing you to compare charts more easily.
Building a chart	It is sometimes desirable to build a chart from a blank template. Click away from any data then use the *Insert* menu to make a blank chart. You can add data using the *Select Data* button.
Named ranges	You can give names to any range of cells. This can help in selecting data for plotting.

PART 2

Using your dataset – summarizing, visualizing and sharing your data

WHAT YOU WILL LEARN IN THIS PART

This part is concerned with how to use your data. This involves using data exploration tools to summarize and visualize your data. You also need to be able to share your data with others.

THE MAIN TOPICS ARE:

- How to tell what kind of dataset you have.
- How to summarize numbers:
  - Using averages.
  - Using measures of dispersion.
  - Using the *Analysis ToolPak*.
- How to look at the shape of your data:
  - Using data frequency and shape statistics.
  - Using histograms.
- How to look for patterns in your data – data mining.
- How to summarize your data numerically:
  - Using correlation matrices.
  - Using contingency tables.
  - Using pivot tables.
- How to choose the appropriate graph for your data.
- How to use sparklines for quick visual summaries.
- How to make different sorts of graph and pivot chart:
  - Using scatter plots for correlation and regression data.
  - Using line plots for time-series data.
  - Using bar charts for association data.
  - Using bar charts for differences data.
    - Adding error bars.
- Using box-and-whisker plots for differences data.

- How to share your data with others:
    - Exporting data in various file formats.
    - Saving pivot tables.
    - Saving charts and sparklines.

4

MAKING SENSE OF YOUR DATA

Getting your data into a sensible arrangement, managing your data and checking over them for errors are important tasks. However, they are simply the forerunner to the most important phase of all, making sense of your data.

There are three main elements to this phase:

- *Exploring your data – data mining*. This is where you look over your data with a view to discovering patterns and trends in your dataset.
- *Summarizing your data*. This is where you distil the information in your dataset in a more concise manner using things like averages.
- *Visualizing your data*. It is difficult to make sense of lots of numbers; graphs make the patterns and trends into a more meaningful display.

These elements are closely interlinked and the spreadsheet tools you'll need will be used in more than one task.

The way you set about exploring and summarizing your data (numerically and visually) depends somewhat on the kind of data that you have. Some datasets will need reorganizing using a pivot table, whilst others will not. In this chapter you'll get an overview of the methods and tools available for you. Subsequent chapters (5–8) deal with the exploration of different kinds of data, according to their general arrangement. Each chapter covers elements of exploring, summarizing and visualizing the data so that you can make the most sense of it.

At some point you'll need to show others what you've done; Chapter 9 covers elements of this, including ways to save your graphs and summary tables.

4.1   TYPES OF DATASET

In a broad sense you can think of data as being in one of three camps:

- Correlation and regression.
- Association.
- Differences.

How you deal with your data depends to a large extent on which camp your data is most allied to. Some datasets may be a combination and contain elements of more than one kind. Once you know what you are looking at you can use the most appropriate tools to help you explore and make sense of your data.

4.1.1 Correlation and regression datasets

Correlation and regression are about finding links between variables. In most cases the variables are numeric and the processes of correlation and regression are intended to determine the strength and nature of the link between two (or more) variables. You can think of correlation as a simple version of regression. In correlation you simply determine the strength (and direction) of the relationship; in regression you define a mathematical relationship between the variables.

In regression you can include several variables and can determine what the effects of these are on the first variable. The first variable, the one influenced by others, is the response variable. The variables doing the influencing are the predictor variables. These predictor variables are usually numeric but you can have categorical variables too. In some kinds of regression your response variable can be categorical too, generally involving two levels (this is called logistic regression).

How you set about exploring the data depends on what sort of variables you've got:

* All numeric – data can be explored in place.
* Some categorical – you'll need to use filters and pivot tables to manage the data.

You'll see details of how you can explore and summarize regression data in Chapter 5. In some cases you can have data that were recorded over time; this can be similar to regression data, and you'll see ways to deal with these data in Chapter 6.

4.1.2 Association datasets

Association analysis involves looking for links between categories of things. Usually you have two sets of categories but there can be other subdivisions. The defining feature of association data is that you do not have replicated data. You have frequency data; each combination of categories contains a single value, the count of all the observations that fell into that combination.

When you record association data you generally record each observation, giving the value for each category involved. There are no numbers! You can construct the contingency table from the raw data; Excel will work out the frequencies for you. You may have data that already have the frequencies included but the upshot is still the same: you need to construct the contingency table to carry out your association analysis.

You'll see how to explore and summarize association data in Chapter 7.

4.1.3 Differences datasets

When you are looking at differences you need to split your dataset into chunks, which you then compare. In the simplest case you have a single response variable and one predictor variable. The response will be numeric and the predictor a factor variable (that is, categorical). You may have a more complicated situation where you have several predictor variables. This does not really affect your analysis, except to make it more complicated! In general you'll need to take your original data and process them into chunks; this will usually mean using a pivot table. You'll see how to explore differences data in Chapter 8.

In some cases your data may involve recording over time, you'll see how to explore time-series data in Chapter 6.

4.2 WAYS TO SUMMARIZE YOUR DATA

It is hard to make sense of a load of numbers. The sensible thing to do is to distil the numbers into a summary. Ideally you should do a visual summary before a numerical one but in practice you have to do the latter before the former. Such summary statistics reduce the original data to a few key values, which help the reader view the data. Summary statistics come in four main sorts:

- Middle values – averages, that is, measures of centrality.
- Variability – whether the values are clustered tightly around the middle or more widespread.
- Replication – how many items you have.
- Distribution – the shape of the data.

Summary statistics are useful but most readers will find a graphic more helpful; these are outlined in Section 4.3. Before that you'll see a brief overview of the useful numerical summary statistics.

4.2.1 Averages

An average is a measure of the centrality of a sample of data, that is, the middle point. There are three main measures of centrality:

- Mean – the arithmetic mean is determined by adding together all the separate values then dividing by the number of items.
- Median – this is based on the ranks of the data. The elements are arranged in size order and the median is the value in the middle.
- Mode – this is the most frequent value in a dataset.

All of these averages can be calculated with Excel formulae, as you'll see next.

Mean
The AVERAGE function will determine the mean value for a sample of data. You simply give the range of cells. This function can also be used directly in pivot tables.

Median
The MEDIAN function will determine the median for a sample of data. You simply give the range of cells. This function is not available in a pivot table but you can of course

make the table showing all the data, set out sample by sample, and then use the function on those.

You can also use the QUARTILE function to determine the median. The function needs the range of cells and the quartile you wish to compute.

```
QUARTILE(range, quartile)
```

The median is the same as the second quartile so you put a 2 in place of the *quartile* parameter. Quartiles are a way of splitting the data into chunks based on the rank of each element. There are four chunks; the boundaries between them are the quartiles. These quartiles are given numbers 0–4:

- Quartile 0 – the minimum value, which you can also compute using MIN.
- Quartile 1 – the upper boundary of the first chunk; this quartile is sometimes called the *lower quartile.*
- Quartile 2 – this is right in the middle, with two chunks below and two chunks above. This is also the median value, which you can determine with the MEDIAN function.
- Quartile 3 – this is sometimes called the *upper quartile* and lies at the boundary of the third and fourth chunks.
- Quartile 4 – the maximum value, which you can also compute using MAX.

Quartiles 1 and 3 are often referred to as inter-quartiles. These inter-quartiles are some-times used to describe the variability of the data (see Section 4.2.2). You cannot use quartiles in a pivot table directly.

Mode

The mode is the number that crops up the most often in a sample of numbers. It can be a useful measure but is not used in statistical analyses. You can compute the mode using the MODE function, which requires simply the range of cells for which you want to obtain the mode.

Sometimes there is more than one modal value. If that is the case then the MODE func-tion will give you the single value that corresponds to the first of the modal values. Thus the order of the data is important. An alternative is the MODE.MULT function (available in Windows Excel 2010 and later). This will compute all the modal values. This function is an *array formula* and is entered slightly differently from the usual function:

1. Highlight several cells; these will hold the results. So if you want to work out up to four modal values highlight four cells.
2. Type =MODE.MULT in the formula bar.
3. Complete the formula by adding the range of cells. You can do this in parentheses by typing the range or using the mouse.
4. Enter the formula in a special way; press Ctrl+Shift+Enter.
5. The result(s) should appear in the highlighted cells.

If there are several modal values you'll see them listed in the order they first appear in the dataset. If you highlight more cells than are required you'll see #N/A as a result. If there is only one modal value then all the cells you highlighted will contain the same value. You'll also notice that the entire formula is enclosed in curly brackets {} when you view it in the formula bar.

> **Note: Array formulae in Excel for Mac computers**
> Array formulae are available on Excel for Mac computers but you need to use Cmd+Shift+Enter.

The mode is not a value that can be used in pivot tables directly.

4.2.2 Variability

If you have a sample of data you can work out the middle value of some kind (mean or median). If all the data are close to the middle value the data has low *variability*. If the data are widely spread out then the data has high variability. Knowing how spread out the data are is important, especially when it comes to statistical analyses.

There are several commonly used measures of variability:

- Standard deviation – this is a kind of average deviation from the mean.
- Variance – this is the standard deviation squared.
- Standard error – this is related to standard deviation but the size of the sample is taken into account. The larger the sample the smaller the standard error.
- Range – this is simply the difference between the maximum and minimum values.
- Inter-quartile range – this is the difference between the first and third quartiles.

These measures of variability can all be calculated easily using Excel formulae.

Standard deviation
The standard deviation is a measure that works out the average distance of each datum to the overall mean. A large value indicates that the data are more spread out than a small value. You'll see a bit more about standard deviation when you look at data distribution (Section 4.2.4).

You can compute the standard deviation using the STDEV function. You simply require the range of cells that contain the values that you want to examine. The standard deviation is also available as a summary statistic in pivot tables.

Variance
The variance is similar to the standard deviation; it is actually the square of it. You can compute the variance using the VAR function. You simply require the range of cells that contain the values that you want to examine. The variance is also available as a summary statistic in pivot tables.

Standard error

The standard error is a measure of variability that is related to standard deviation (and also variance). In effect the standard error is a measure of how good your estimate of the mean value is. In practice you cannot measure this so it is estimated by using the standard deviation and the sample size:

```
Std. Error = Std. Deviation ÷ √Sample size
```

There is no function to compute standard error directly but you can use the COUNT function to get the sample size, the SQRT function to get the square root and the STDEV function for the standard deviation. Needless to say, this is not a measure that you can use in a pivot table.

Standard error is often used in statistical analyses. It is common to use the standard error for error bars on graphs, rather than standard deviation, since the measure is related to sample size. You'll see error bars used (with standard error) especially in Section 8.2, when you'll look at bar charts and box-and-whisker plots to visualize differences data.

Range

The range is simply the difference between the maximum and minimum values. You can use the MAX and MIN functions to get the maximum and minimum respectively. There is no function that takes the range as an argument, so you'll need to subtract one from the other.

The maximum and minimum are available to use in pivot tables but you cannot compute the range directly in a pivot table.

You can also use the LARGE and SMALL functions to determine the maximum and minimum values. These require two parameters, the range of cells and a number corresponding to the value you want. So, you enter a 1 (one) to get the first largest or first smallest. In general it is simpler to use MAX and MIN functions.

Inter-quartile range

The inter-quartile range is analogous to the standard deviation or standard error. Essentially you are getting the range of the data but chopping off extreme values from the highest and lowest ends. You can use the QUARTILE function to work out the first and third quartiles:

```
QUARTILE(range, 3) - QUARTILE(range, 1)
```

The inter-quartile range is the third quartile minus the first quartile; there is no function to work this out. Needless to say, this cannot be done in a pivot table directly.

4.2.3 Replication

Replication refers to how many items you have in each sample. Sometimes this is known as the number of observations (or number of records). You can use the COUNT function to count the number of items (non-blank cells) in a range of data.

The number of replicates is used in calculations of standard error:

```
STDEV(range) / SQRT(COUNT(range))
```

You can use the count of items in pivot tables.

4.2.4   Data distribution

The distribution of data refers to the shape. You can examine the shape of the data using a frequency table. This involves looking at each value and working out how many times it occurs in the sample.

What you expect is that values near to the average will be higher in frequency to the values at the extremes. If the values are more or less symmetrically arranged around the middle then your data can be described as normal (also called parametric). If there are rather more values to one side or the other of the middle then your data are skewed (also called non-parametric).

The shape of the data affects the sort of summary statistic that is most appropriate (Table 4.1) and the kind of statistical analysis that you can carry out.

Table 4.1 The distribution of your data determines which summary statistics are most appropriate.

	Parametric data	Non-parametric data
Average	Mean	Median
Variability	Standard deviation	Range
	Standard error	Inter-quartile range

Usually you'll visualize the data distribution in the form of a histogram, which is a sort of bar chart where each bar represents a range of values. The height of the bars is the frequency of data occurring in each data range. You'll see more about histograms shortly (Section 4.3.1). Before you can make a histogram you need to determine the frequency of the data in various data ranges.

> **Note: Excel bar charts**
> Excel column charts are what most people call bar charts. Excel bar charts display the bars horizontally. The difference in nomenclature is presumably a device to make it easier to label the chart buttons sensibly on the *Insert* menu.

Data frequency

If you have a sample of data you can split it up into equal-sized chunks, called *data bins* or simply *bins*. To begin with you'll need to work out how many bins you want. This is something of a black art: if you have too few then you won't see a decent pattern, as everything will be packed into the bins. If you have too many bins the items can be too spread out, also making it hard to see a pattern.

It is generally useful to have an odd number, because you'll definitely have a central bin, but it is not essential. I would start with seven bins and work from there. Once you have a starting point you can determine the values that will occupy the bins and then calculate the frequencies using the FREQUENCY function. The whole process can be summarized thus:

1. Calculate the range of your sample using the MAX and MIN functions.
2. Decide how many bins you want. Then calculate an interval between bins by dividing the range by the number of bins–1.
3. The first bin can be the minimum value from your sample so compute that using the MIN function.
4. Subsequent bins will be the value of the previous bin plus the interval you calculated in step 2. You'll need the $ to fix the cell reference in a simple formula.
5. Fill down the formula you used in step 4 to determine values for all the bins you need.
6. Highlight the cells adjacent to your bins, these will hold the frequencies. The cells do not actually need to be adjacent but it is easier to see what you've done if they are.
7. Now in the function bar type =FREQUENCY to start the computation of the frequencies.
8. The function needs the range of cells so open a parenthesis and use the mouse to highlight the cell range. Type a comma and then select the range of values corresponding to the bins.
9. Close the parenthesis to complete the function. Complete the calculations by pressing Ctrl+Shift+Enter, as the FREQUENCY function is an array function (a special sort of function that fills its results into an array of cells).
10. The frequencies should be displayed alongside the bins. You can use the SUM function to total the frequencies, to check that you have the same number as the number of items in the sample.

In step 3 you computed the first bin to correspond to the minimum value of the sample. Obviously you want a value that is at least the minimum (otherwise you'd have zero frequency) but you could make a larger value if you prefer.

Tip: Easily interpreted bin values
Histograms can be easier to understand if the bin values are easily understood numbers. You can tweak the starting point and interval to produce more easily interpreted values. Integers are best, then halves (0.5) and so on.

If you alter any of the values in the bin range the result will update immediately but if you add more cells you'll need to repeat the process of calculating the frequencies.

The frequency table can be useful and interesting but it is a lot more useful when used in conjunction with a histogram, which you'll see later (Section 4.3.1). It is easier to see the pattern visually than in the form of a table.

> **Note: Array results**
> If you've used an array function to work out some results, such as data frequency, you cannot insert cells into the array. You'll have to delete the formula from the cells first.

Categorizing the shape of the distribution

You can look at a table of frequencies, or better still a histogram, and get an impression of the shape. If the shape appears to be symmetrical you are likely to have normal (parametric) distribution. If the largest frequency appears towards one end you've got non-parametric distribution. The distribution of the data will affect the kind of statistical analysis you can carry out so it is important to know.

There are a couple of statistics that can help you characterize the shape of the distribution:

- Skewness – this is a measure of how central the average is in your data sample.
- Kurtosis – this is a measure of how pointy your histogram is.

Taken together with a histogram you gain a good impression of the distribution of your data.

Skewness

Skewness is a measure of how symmetrical your data are. The nearer the value is to zero the more symmetrical. If you have a positive value your histogram will have its tallest bar(s) towards the left; the longer tail of the histogram will be to the right. If you have a negative value then the histogram will have its tallest bar(s) to the right and the long tail will be to the left.

You can calculate skewness with the SKEW function, which requires the range of cells corresponding to the data sample you wish to explore.

Kurtosis

The kurtosis of a data sample is a measure of how pointy it is. It is easier to see in conjunction with a histogram. If the kurtosis is a positive value the histogram is more pointed. If the kurtosis is negative then you have a flatter distribution. Think of it like this: really high positive values make the histogram look like a skyscraper in a village (the only tall building around). As kurtosis gets smaller the buildings become more even in size.

You can calculate kurtosis using the KURT function, which requires the range of cells corresponding to the data sample you wish to explore.

In the following exercise you can have a go at calculating some distribution statistics for yourself.

Have a Go: Compute data distribution statistics

You'll need the *Pottery.xlsx* file for this exercise. The data show chemical composition of various pottery sherds from several sites. There are five columns relating to the chemicals and one for the site name.

↘ Go to the website for support material.

1. Open the *Pottery.xlsx* spreadsheet. The data are in the workbook entitled *Data*. You'll start by looking at the distribution of the *Al* column (the concentration of aluminium). Go to cell J1 and type a heading, `Al distrib` will do.

2. Now go to cell J2 and type `Range`, then in cell J3 type `Interval`. In cell J4 type `Bins`. Now go to cell K4 and type `Freq` as a heading. You'll use these headings to help keep track of the statistics you'll compute.

3. Go to cell K2 and type a formula to give the range of the *Al* data variable: `=MAX(C2:C27) - MIN(C2:C27)`.

4. Go to cell K3 and type a formula to work out the interval for seven bins, `=K2/7`.

5. In cell J5 type a formula to work out the smallest value in the *Al* variable: `=MIN(C2:C27)`.

6. In cell J6 type a formula to work out the next bin value; this will be the previous value plus the interval: `=J5+ $K$3`. Don't forget that you'll need to fix the cell reference for the interval using $.

7. Now copy the formula in cell J6 down to cell J11 so that you end up with seven values relating to the bins.

8. Use the mouse to highlight the cells next to the bins, that is, cells K5:K11.

9. Click in the function bar or simply start typing. Type a formula to work out the frequencies: `=FREQUENCY(C2:C27, J5:J11)`. The first cell range is the Al variable and the second is the bin values from steps 5–6. Complete the formula by pressing Ctrl+Shift+Enter on the keyboard.

10. Now go to cell J13 and type `Skew` as a heading. Under this (cell J14) type `Kurtosis`.

11. In cell K13 type a formula to work out the skewness of the *Al* variable: `=SKEW(C2:C27)`.

12. In cell K14 type a formula to work out the kurtosis of the *Al* variable: `=KURT(C2:C27)`.

You now have a frequency table and some additional statistics for the *Al* variable. Try using named ranges (see Section 3.3.2), which will allow you to type the names into formulae instead of using the mouse. This will allow you to explore the frequencies of the remaining variables rather more easily.

Keep the spreadsheet open for the time being, as you'll have a go at making a histogram shortly. Alternatively save the spreadsheet as a new file, to keep your calculations intact.

The summary statistics you've seen so far have all involved regular Excel formulae. It is also possible to get summary statistics using an add-in for Excel, available for Windows versions. This is the *Analysis ToolPak*, which you'll see next.

4.2.5   Using the Analysis ToolPak

The *Analysis ToolPak* is a Microsoft add-in for Excel. It is available for all Windows versions but not for Mac (except quite old versions). There is no direct equivalent in Open Office.

The *Analysis ToolPak* contains a range of tools for carrying out statistical analyses, including some of the summary statistics you've already seen. Shortly you'll see how to use it to create a histogram (Section 4.3). Later you'll see how it can be used to explore regression data (Section 5.1). First of all you'll need to see how to install it.

Installing the Analysis ToolPak
The *Analysis ToolPak* is not usually activated by default. You need to actively turn it on using Excel options. To activate the *Analysis ToolPak* you need to follow these steps:

1. Click the *File* tab on the ribbon and select *Options*.
2. Choose *Add-Ins* from the options on the left.
3. You will see various sections including those entitled *Active Application Add-ins* and *Inactive Application Add-ins*.
4. If you see the *Analysis ToolPak* listed in the *Active Application Add-ins* section then you don't need to go any further.
5. At the bottom of the screen there should be a box labelled *Manage*. There is a *Go* button to the right of that. Make sure that *Excel Add-ins* is selected and then click the *Go* button to open the *Add-Ins* dialog box (Figure 4.1).

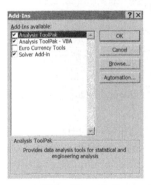

Figure 4.1 The *Add-Ins* dialog box is accessed from *File > Options > Add-Ins* using the *Manage* section.

6. Use the mouse to tick the add-ins you want to enable (the *Analysis ToolPak*) and click *OK* once you are done.
7. The add-ins appear immediately in the *Data* menu, as *Data Analysis* in the *Analysis* section (you don't need to restart Excel).

There are various tasks you can accomplish using the *Analysis ToolPak*; producing a range of summary statistics is one, which you'll see next.

Summary statistics using the Analysis ToolPak

You can use the *Analysis ToolPak* add-in to give you a range of summary statistics. Once the add-in is enabled (see previous section) you can use it from the *Analysis* section of the *Data* menu. To summarize a sample you follow these steps:

1. Open the spreadsheet containing the data you want to summarize.
2. Navigate to the worksheet containing the data and then click the *Data > Data Analysis* button. This opens the *Data Analysis* dialog box (Figure 4.2).

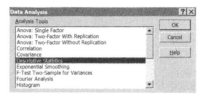

Figure 4.2 The *Data Analysis* dialog contains the options available in the *Analysis ToolPak*.

3. Select the *Descriptive Statistics* option then click *OK*. The *Descriptive Statistics* dialog box now opens, allowing you to select the data and where to place the output (Figure 4.3).

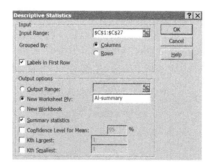

Figure 4.3 The *Descriptive Statistics* dialog box allows you to select input data and where to place the results.

4. Click the *Input Range* box and select the data; you can include a title cell.
5. Usually you'll have data arranged by column but check that the option *Grouped by Column* is selected. If your data includes a title then click the check box labelled *Labels in First Row*.
6. Select a location for the output; a *New Worksheet Ply* is most sensible. You can type a name for the worksheet if you like.
7. Now check the box entitled *Summary statistics*.
8. Click the *OK* button to carry out the computations. The results will be placed in a new worksheet or wherever you selected in step 6.

The summary statistics include all those you've seen so far and the *sum*. If you selected more than one variable they are all listed (in adjacent pairs of columns), so it is a quick way to get a numerical overview of your sample data.

Summary statistics in the form of numbers are useful but visual summaries are more powerful; a graph is interpreted more readily than a table of figures. You'll see some of the ways you can represent your data visually in the following section.

4.3 WAYS TO VISUALIZE YOUR DATA

Graphs are a powerful way to summarize your data. A good graph can speak volumes so it is worth spending some time to get the best you can. You'll see how to save and share your graphs later (Section 9.3), and in Chapters 5–8 you'll see how to use charts to help you explore various sorts of dataset. First you'll look at the sorts of graph available and get an idea of what each graph can do for you.

4.3.1 Types of graph

Excel can produce an extensive range of graphs. Some kinds of graph are easy to produce and others need rather more work. The graphs are all available via the *Charts* section of the *Insert* menu. Some graphs can also be produced as pivot charts, which are closely linked to pivot tables. You met pivot charts when error checking (Section 3.3) and you'll see them again soon (Section 4.3.2).

It is useful to know to what uses you can put the various chart types. In Table 4.2 you can see a list of graphs and the main uses they are suitable for. Note that not all of the types of chart that Excel is able to produce are listed. Some kinds of chart are rather specialist and not very useful in a scientific context.

The graphs listed in Table 4.2 are the most useful of the kinds that are available in Excel. Note however that some are rather more challenging to produce. The dot chart for example you've met earlier in the context of error checking (Section 3.3). Excel does not have a built-in dot chart and you need to convert a line plot (which you can save as a template).

The box-and-whisker plot (also called simply boxplot or box-whisker) is potentially very helpful, as it can convey a lot of information in a concise manner, but it is tricky to produce, as you need to use a stock chart from the *Custom* section. You'll see more about box plots when you come to explore differences data (Section 8.2.4).

Pie charts remain popular in spite of not being necessarily the best choice to display information. The human eye is not good at interpreting angular data and you can convey the same information as a bar chart more effectively.

The types of graph in Table 4.2 fall into two groups. Most of the graphs are used to summarize the data by showing averages of some kind; the histogram is the odd one out as it displays the distribution of the data.

Table 4.2 Types of chart and some of their uses.

Chart type	Main uses
Bar chart Column chart	Showing data in categories. Association data. Differences data.
Box-and-whisker plot (also called boxplot, box-whisker)	Showing data in categories. Differences data.
Scatter plot	Comparing numeric variables. Correlation and regression data.
Line plot	Showing how a numeric variable changes over time. Time-related data.
Dot chart	Showing data in categories. Error checking samples.
Pie chart	Showing compositional data in categories. Association data.
Histogram	Showing data distribution. Frequency data in categories.

Summary graphs

Most of the charts in Table 4.2 are used for summarizing data in some way. Each graph is suitable for certain kinds of data; each type of data can also be summarized by certain types of chart:

- Correlation and regression data – scatter plots are the most useful. They display numeric data on both x and y axes (see Section 5.2.2).
- Association data – bar and column charts are most useful although dot charts can be pressed into service. The data are categorical and bar charts (and column charts) display values split into categories (see Section 7.2).
- Differences data – bar charts, column charts and box plots are most useful although dot charts can be pressed into service. You want to be able to display average values (and variability) split by sample groups and this is exactly what you can do using these graphs (see Section 8.2).
- Time-related data – line plots are most useful but scatter plots or bar charts are sometimes used. A line plot shows numerical values on the y-axis and categories on the x-axis. The points are generally linked by a line (hence the name), which implies a link between data points. In time-related data this is relevant; you can look at patterns of change by following the line (see Section 6.2). A scatter plot is sometimes relevant, especially if the time intervals are not equal.

Data distribution: the histogram

The histogram is a special form of bar chart. Excel calls a categorical chart with vertical bars a column chart and one with horizontal bars a bar chart. In general parlance a bar chart is the name used for both kinds. You can make a histogram using either a column or bar chart from the *Insert* menu, although it is more usual to make a column chart (it is only a convention that vertical bars are used!). You can also use the *Analysis ToolPak* to calculate the frequencies and create a chart.

In any event you need to split the sample into chunks (bins). If you use the regular chart approach you'll have to determine the frequency for each bin. If you use the *Analysis ToolPak* you can skip this part and use the bins only.

A histogram shows the range of data on the *x*-axis as a series of bins. In a true histogram the *x*-axis is displayed as a continuous axis and the divisions between the bins are labelled with the value. This highlights the fact that the data range is continuous. In Excel the histogram is not quite like this; each bar is labelled with the maximum value in that bin, with the labels between the divisions.

The *y*-axis displays the frequency of data for each bin. In a true histogram the bars do not have gaps between them, again highlighting the fact that the *x*-axis is showing a continuous range of values.

Histogram from regular charts

If you want to construct a histogram using regular charts (bar or column) you'll need to determine how to split the sample into bins. You'll then need to calculate the frequency of the data in each of the bins. You saw how to do this using the FREQUENCY function earlier (Section 4.2.4).

Once you have your bins and the frequencies you can make the histogram. Follow these steps to make the chart:

1. You need to make a blank chart so click anywhere in a worksheet that is not in or adjacent to any data. A separate worksheet is ideal.
2. Click the *Insert > Column* button and select the *2-D Clustered column*, which is the plain option. The blank chart is created immediately.
3. The *Chart Tools* menus should be active but if not click once on the blank chart. Then click the *Design > Select* data button.
4. Click the *Add* button (in *Legend Entries (Series)* section) to open the *Edit Series* dialog box.
5. Type a name for the series in the *Series name* box, or point to a cell containing a label or simply leave it blank.
6. Now click in the *Series values* box and delete anything that appears. Use the mouse to select the frequency values you calculated earlier. Select only the numbers and not any labels you may have made. Click *OK* to return to the *Select Data Source* dialog box.
7. The histogram is now constructed but the *x*-axis labels are plain index numbers. Click the *Edit* button in the *Horizontal (Category) Axis Labels* section to open the *Axis Labels* dialog box.

8. Use the mouse to select the bin values you made earlier; you want only the numbers so do **not** include any labels. Click *OK* to return to the *Select Data Source* dialog box.
9. The *x*-axis labels should now reflect the bins so you can click *OK* to return to the worksheet and the histogram.
10. Click once on a bar. Then open the *Format Data Series* dialog by right-clicking and selecting *Format Data Series*. You can also use the *Chart Tools > Format > Format Selection* button.
11. Select the *Series Options* section, and then use the slider to reduce the *Gap Width* to zero.
12. Now go to the *Border Color* section and make the border a *Solid line* of a contrasting colour (black or white is good). Click *Close* to return to the histogram.
13. You can now use the *Chart Tools* menus to help you tweak the chart; the legend is not required for example. You may want to add axis titles (see Section 9.3.2 for details of things to consider when sharing charts).

The histogram bin labels usually take their format from the data so to reduce the number of decimal places displayed on the chart you can format the bins themselves. A completed histogram should give you an impression of the data distribution (Figure 4.4).

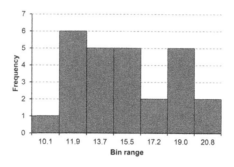

Figure 4.4 A completed histogram, edited from a column chart.

You can work out the range of values covered by each bin from the axis labels. In Figure 4.4 the first bin is labelled 10.1; this covers a range from 10.1 (inclusive) downwards. The next bar is labelled 11.9 and so contains items that are larger than the previous bar (10.1) and up to (and including) 11.9. The third bar contains items greater than 11.9 and up to 13.7 (inclusive). Sometimes you'll see a label *More*, which indicates no absolute top value.

You can alter the chart type using the *Chart Tools > Design > Change Chart Type* button. The standard column chart is most appropriate but you may find uses for other types!

Histogram using the Analysis ToolPak

The *Analysis ToolPak* can take some of the effort out of making a histogram. To start with you do not need to use the FREQUENCY function; the frequencies are calculated for you. Secondly, the chart itself is created quickly and you only need to edit it to tidy it up afterwards.

You don't need a blank worksheet to begin with; all you need is a spreadsheet containing your data and the cells with the bins. The *Analysis ToolPak* does the rest. The process is:

1.   Click the *Data > Data Analysis* button to start the *Analysis ToolPak.*
2.   Select the *Histogram* option to open the *Histogram* dialog box (Figure 4.5).

Figure 4.5 The *Histogram* dialog in from the *Analysis ToolPak.*

3.   Click in the *Input Range* box and then use the mouse to highlight the data. There is no need to include the header, just the values.
4.   Now click in the *Bin Range* box and select the data cells that contain the bin values.
5.   Check the option to place the output in a *New Worksheet Ply*; you can name the worksheet if you like.
6.   Tick the box labelled *Chart Output*. This makes a chart as well as computing the frequencies. If you don't tick this you'll get a table of frequencies only. Click *OK* to complete the operation and make the histogram.
7.   Click once on the chart to activate the *Chart Tools* menus. You should also see the data selection borders appear in the table in the worksheet (it has headings *Bin* and *Frequency*).
8.   You'll see that the last entry in the table is labelled *More.* This is a catch-all category that allows for cases when you've not made your bin range large enough. Hover over the bottom right corner of the *Frequency* column until the cursor alters to a two-headed arrow. Then click and drag to reduce the number of rows so that the *More* row is not included.
9.   Use the *Chart Tools* menus to edit the chart. You need to make the bars wider so that there are no gaps (see the preceding section) and remove the legend at the very least.

Tip: Histogram bars
Instead of having touching bars (that is, with zero gap) try setting the gap width to a small value (2%). This allows the individual bars to be seen separately and saves having to reformat the borders.

Earlier you computed some frequencies from sample data (Section 4.2.4). In the following exercise you can have a go at using these frequencies to make a histogram. You can also use the same data with the *Analysis ToolPak* to make a histogram.

Have a Go: Construct histograms

You'll need the *Pottery.xlsx* spreadsheet for this exercise. If you kept the frequency calculations from the previous exercise then use those. If you did not keep the frequency calculations then repeat the previous exercise to make the frequency table for the *Al* variable.

↘ Go to the website for support material.

1. Make a new worksheet for a histogram from the frequency data for the *Al* data variable.
2. Click the *Insert > Column* button and then select the regular *2-D Column* chart.
3. Click once on the blank chart and then use the *Design > Select Data* button.
4. Click the *Add* button then click in the *Series* values box. Select the frequency data you calculated then click *OK*.
5. Now click the *Edit* button on the right, in the *Horizontal (Category) Axis Labels* section.
6. Use the mouse to select the values for the bins; these will form the *x*-axis. Click *OK* to complete this operation. Click *OK* again and the basic histogram is made.
7. Right-click on a bar in the chart and select *Format Data Series*. Then from the *Series Options* section alter the *Gap width* to a very small value (2% is about right). Click *Close* to return to the chart.
8. Edit the chart as you like to make it nicer; you might want to add axis titles and remove the legend for example.
9. Now return to the *Data* worksheet to make a histogram using the *Analysis ToolPak*. Click the *Data > Data Analysis* button to open the *Analysis ToolPak*.
10. Select *Histogram* and click *OK*.
11. From the *Histogram* dialog box select the data in column C for the *Input Range* (only the values, not the heading). Select the bin values for the *Bin Range* section. Make sure the output is set to *New Worksheet Ply* and check the box labelled *Chart Output*. Click *OK* to make the histogram.
12. Click once on the chart then use the data selection border to remove the *More* row from the chart. Hover over the bottom right corner and click-drag the mouse up one row to achieve this.
13. Alter the bar width and edit as necessary (steps 7–8). You completed histogram should resemble Figure 4.4.

Try making histograms of other variables in the data. You'll need to work out appropriate bins to do this. There are a few short cuts you can take but Excel is not really set up for this kind of analysis so you'll have to be creative!

Making histograms is an important task, as you need to determine the shape of the data distribution in order to determine the appropriate statistical test(s). However, it is not a task that can be streamlined very easily since each sample has a different bin range. If you make a new worksheet you should be able to arrange things so that you have calculations for each variable in a separate column; you want the bins and the frequencies. Once you have one variable calculated you can copy and paste to new columns. If you then make a histogram from a blank column chart you can use the dynamic selection borders to switch between variables fairly easily.

Tip: Semi-automatic histogram
If you leave the *Bin Range* box blank when you use the *Histogram* part of the *Analysis ToolPak* add-in Excel works out the bins for you.

Note: Histograms and multiple data series
You can't add extra samples to your histograms unless you have the same bin ranges for all the samples you want to chart.

If you have a sample that is split into different levels by another variable (a predictor or grouping variable) you'll have to use an index variable and a pivot table to rearrange the data into separate columns. You can then explore the distribution of each separate group.

4.3.2 Pivot charts

A pivot chart is a graph that is linked closely with a *pivot table*. You met the pivot chart earlier when looking at error checking (Section 3.3). Pivot charts are especially useful for summarizing sample data, allowing you to group your data by different variables and use the mean.

You can't use all types of graph with a pivot chart; generally you'll use bar charts and line plots. You cannot use scatter, bubble or stock charts.

The advantages of the pivot chart are:

- They are easily modified.
- You can apply filters.

This allows you great flexibility in exploring your data because you can present chunks of your data in various ways quickly and easily. Pivot charts are particularly useful for:

- Differences data.
- Association data.
- Time-series data.

Pivot charts are less useful for correlation and regression data, where you'll usually need scatter plots.

You can make a pivot chart in two ways:

- Direct – use the *Insert >PivotTable> PivotChart* button and construct the chart in a similar way to a pivot table (see Figure 3.21).
- Indirect – use the *Insert >PivotTable* button to make a pivot table then use the *PivotTable Tools > Options > PivotChart* button to create your chart.

If you use the direct method the default is a column chart; you can use the *Design > Change Chart Type* button to alter the chart type. If you use the indirect method you'll see the *Insert Chart* dialog box when you use the *PivotChart* button; you can select the chart type from that.

You'll see pivot charts used in the Chapters 5–8 for exploring different kinds of dataset.

4.3.3 Sparklines

Sparklines are a kind of mini-graph and are available in Excel versions from 2010 onwards. You can find the Sparklines tools on the *Insert* menu of the ribbon (Figure 4.6).

Figure 4.6 *Sparklines* tools are in a section of the Insert menu on the ribbon.

There is no equivalent in Open Office; it is only later versions of Excel that have this facility. The sparkline comes in two main forms, a column chart and a line plot. There is also a Win/Loss type but this is generally less useful, it simply shows positive or negative values.

A sparkline fits within a single cell and it is possible to enter data in the same cell; the sparkline acts as a background.

In order to add sparklines you need the following steps:

1. From the *Insert > Sparklines* menu click the button that represents the type of chart you require, mostly this will be *Line* or *Column*.
2. The *Create Sparklines* dialog box appears and you should select the source data for the *Data Range* field (Figure 4.7).

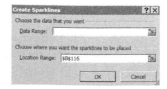

Figure 4.7 The *Create Sparklines* dialog box allows you to select source data and the destination for the completed sparklines.

3. Select the appropriate data (values only, do not include any headers). If you want more than one column (or row) then include all as appropriate.

4. Now select the *Location Range* for the sparkline. Usually the *Location Range* box contains the cell that was active when you started the process, simply select the target cells.

5. Click *OK* to create the basic sparkline. If you selected too many or too few cells you will get a warning message and will have to reselect an appropriate range of cells.

6. The *Sparkline Tools Design* menu should now be active on the ribbon. You use this to alter the appearance of your sparkline.

The *Sparkline Design* menu offers you various ways to edit your *Sparklines*. It can be useful to alter the colours of the high and low points for example (Figure 4.8).

Figure 4.8 The *Style* section of the *Sparkline Tools > Design* menu allows you to alter the colours of various sparkline components.

Sparklines can be useful in allowing you a quick view of samples. This can help to spot patterns and trends quickly (Figure 4.9).

	A	B	C	D	E
1	Obs	Ozone	Radiation Temp		Wind
110	109	14	191	75	14.3
111	110	18	131	76	8.0
112	111	20	223	68	11.5
113					
114					

Figure 4.9 Completed sparklines can help spot patterns in datasets. Here, each sparkline summarizes a separate column variable.

You'll see sparklines used in the Chapters 5–8 as you look at exploring different kinds of dataset.

> **Note: Sparklines in rows or columns**
> If your sparklines form a row of cells the data in each summarizes the column it is in. If your sparklines form a column of cells the data in each summarizes the row it is in.

4.4 EXERCISES

1. What are the main measures that you need to summarize a data sample?
2. A histogram is a kind of ____ chart that allows you to visualize the ____ of a data sample.
3. A line plot is most useful for looking at correlation and regression data. TRUE or FALSE?
4. Which of the following charts **cannot** be constructed using a pivot chart?

   A) Bar chart
   B) Scatter plot
   C) Pie chart
   D) Dot chart
   E) Column chart

5. Look at the *jackal.xlsx* data file. The data represent two samples of mandible length (in mm) for male and female golden jackals. Have a look at the skewness and kurtosis of the two samples. How do these relate to the difference between the mean and median values?

↘ Go to the website for support material.

The answers to these exercises can be found in Appendix 1.

4.5 SUMMARY

Topic	Key points
Types of dataset	Different sorts of data require different approaches to dealing with them. The three basic kinds are:
	Correlation and regression – you are looking for links between variables in terms of the strength and nature of the relationship.
	Association – you are looking for links between categories of things.
	Differences – you are looking for differences between groups of data (samples).
Summarizing samples	A numerical summary makes a sample of values more easily understood. There are four kinds:
	Middle values – averages.
	Variability – how tightly clustered around the middle the data are.
	Replication – how many observations are in each sample group.
	Distribution – the shape of the data.

Topic	Key points
Averages	An average is a numerical way of defining the middle point of a sample of numbers (a measure of centrality). There are three main averages:
	Mean – calculated by adding together the values and dividing by the number of observations. The AVERAGE function can do this.
	Median – the value in the middle if the data were arranged in rank order. The MEDIAN function can do this. The QUARTLIE function can also calculate the median (the second quartile is the median).
	Mode – the most frequent value (there can be more than one) in a sample. The MODE function can do this.
Variability	The variability is a way of defining the dispersion of a sample, that is, how spread out the data are from the middle. There are several measures:
	Standard deviation – the average deflection from the mean. The STDEV function can do this.
	Variance – related to the standard deviation (variance is standard deviation squared). The VAR function can do this.
	Standard error – this is related to the standard deviation and the sample size (standard deviation divided by the square root of the sample size). You can use the STDEV function to calculate this.
	Range – the difference between the maximum and minimum values. You can use the MAX and MIN functions to calculate this.
	Inter-quartile range – the sample is split into four chunks, based on their ranks (the quartiles are the boundaries between chunks). The inter-quartile range is the difference between the third and first quartiles. You can use the QUARTILE function to help you determine this.
Replication	This is simply the number of observations (records) in a sample. You can use the COUNT function to help you determine this.
Data distribution	The shape of the data can be determined by looking at the frequency of values in a sample. Most often the range of values is split into chunks, called bins.
	The FREQUENCY function can help determine the frequency of values in various bins. This is a function that sends its result(s) to an array of cells.
	The shape of the data will determine which summary statistics are most appropriate. If the shape is normal (also called parametric), with a symmetrical shape, you can use mean and standard deviation. If the shape is skewed (also called non-parametric) you need the median and inter-quartile range.
	You can show the shape of a sample in various ways:
	Skewness – a numerical measure of how central the average is in the sample. The SKEW function can compute this.
	Kurtosis – a measure of how pointy the distribution is. The KURT function can calculate this.
	Histogram – a kind of bar chart that shows the frequency of observations in the various bin ranges.

Topic	Key points
Analysis ToolPak	The *Analysis ToolPak* is an Excel add-in that can carry out a range of useful tasks including: summary statistics and histograms.
Types of graph	Different type of graph are suited to particular uses: Scatter plots – for correlation and regression data. Bar and column charts – for association and differences. Box-and-whisker plots (boxplot) – for differences. Line plots – for time-related data. Dot charts – for showing data in categories. Pie charts – for compositional data.
Histogram	A histogram is a kind of bar chart that shows the frequency of data in various bin ranges. You can construct one from a frequency table or by using the *Analysis ToolPak*. The histogram helps you determine the shape of the data and therefore the most appropriate summary statistics.
Pivot charts	Pivot charts are closely allied to pivot tables and are constructed using a pivot table as a starting point. Pivot charts allow you to explore your data in various ways quickly and easily but you can only make some kinds of chart. You cannot make a scatter plot using a pivot chart.
Sparklines	Sparklines are simple charts that are completely contained in a single spreadsheet cell. You can make simple line plots and bar charts. Sparklines are quick to produce and are useful as a brief overview – especially if you use them in conjunction with a data table.

5

EXPLORING REGRESSION DATA

When you have regression data you are mostly looking for relationships between numeric variables. You may also have grouping variables, which will be categorical, that is they show different levels. Correlation is a simpler form of regression and is solely concerned with the strength (and direction) of the link between two numeric variables.

If you have grouping variables you are most likely looking at differences between the groups defined by these variables. However, some forms of regression are possible and you may want to explore the different regression relationships between the groups.

In this chapter you'll see how to explore regression datasets (and by implication, correlation data) that include all-numeric data and grouping variables. These grouping variables make it a little more difficult in Excel, but that makes it more worthwhile looking at!

There are two main ways to explore your data:

- Numerically – using correlation matrices.
- Visually – using scatter plots and other kinds of chart.

You'll see ways to visualize the relationships between the variables in Section 5.2 but first you'll look at numerical methods of exploration.

5.1 FINDING CORRELATIONS IN YOUR DATASET

The starting point for most regression analysis is to look at the relationships between the variables pair by pair. In other words you look at simple correlations. Usually you'll have one or more response variables and one or more predictor variables. When you look at the correlations between the various variables you assemble them into a correlation matrix. This allows you to see all the relationships in one compact table. The advantages of a correlation matrix (as compared to a list for example) are:

- You can spot the response variables that have the greatest effect on the predictors.
- You can spot response variables that are closely related to one another.

These are important things to know, and the correlation matrix is the most compact way to display the information.

The approach you take depends upon the data. If your data are all numeric things are fairly easy (Section 5.1.1). If you have grouping variables you need to put in a bit more effort, as you'll see later (Section 5.1.2).

5.1.1 Correlation matrices using all-numeric data

If your variables are all numeric you have two main options for making a correlation matrix:

- Use the *Analysis ToolPak* to compute a correlation matrix.
- Use the CORREL function to compute the link between variables.

The CORREL function gives you fine control over the result but it can be tedious to use the function repeatedly, especially with large datasets. The *Analysis ToolPak* has a *Correlation* routine that will produce a correlation matrix from the columns you select. You'll see this ideal labour-saving tool now.

Correlation using the Analysis ToolPak
The *Analysis ToolPak* is available from the *Analysis* section of the *Data* menu on the ribbon. It is an Excel add-in, which is available for Windows versions of Excel (some older Mac versions too but not Open Office). You saw the *Analysis ToolPak* earlier (Section 4.2.5), when you looked at making histograms to look at data distribution.

Using the *Analysis ToolPak* for correlation is quite straightforward:

1. Open the spreadsheet containing the data you want to explore.
2. Click the *Data > Data Analysis* button to open the *Data Analysis* dialog box (see Figure 4.2).
3. Select the *Correlation* option and click *OK* to open the *Correlation* dialog box (Figure 5.1).

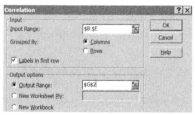

Figure 5.1 The *Correlation* dialog box from the *Analysis ToolPak* allows you to select data to use in a correlation matrix.

4. Click in the *Input Range* box and select the data you want to correlate. You can select entire columns by clicking the column headers. The block of data must be contiguous.
5. If you included the variable names from the first row (which is a good idea) tick the box labelled *Labels in first row*.

6. Click an option in the *Output options* section. Generally *New Worksheet Ply* is a good idea (you can select a name for the worksheet if you like). If you do want to place the output in the same worksheet then click in the *Output Range* box and then click the cell where the top-left of the result is to be placed.

7. Click *OK* to carry out the computations; the correlation matrix will be placed in your spreadsheet at the location you selected (Figure 5.2).

	G	H	I	J	K
		Ozone	Radiation	Temp	Wind
Ozone		1			
Radiation		0.348342	1		
Temp		0.698541	0.294088	1	
Wind		-0.61295	-0.12737	-0.49715	1

Figure 5.2 A correlation matrix produced by the *Analysis ToolPak*.

Your correlation matrix contains the pair-by-pair correlations for all the variables you selected. Hopefully you'll have included the variable names. You cannot alter the results (look at the table and you'll see there are no formulae), but you can reformat to a sensible number of decimal places. Because the table is fixed it will not update if the source data changes; you will need to run the *Analysis ToolPak* again.

The main limitation of the *Analysis ToolPak* is that you cannot select non-adjacent columns; the block of data must be contiguous. This should not present too many problems because you ought to have arranged your data with all the variables in a block anyhow (see Section 2.3.2 on rearranging columns).

If you do want to be more selective you can use the CORREL function to explore correlations, as you'll see next. If you have grouping variables then you simply can't use the *Analysis ToolPak* and CORREL is the only way to go (see Section 5.1.2 later).

Correlation using the CORREL function

The CORREL function computes the linear correlation (Pearson product moment) between two numeric variables. It is therefore possible to calculate the correlation between all the variables pair by pair.

```
CORREL(sample1, sample2)
```

You simply provide the function with the cell ranges that correspond to the two samples you wish to correlate. The samples should be the same size: if they aren't you should get an error, #N/A.

You can select the range of cells using the mouse or by making named ranges (recall Section 3.3.2). Using the mouse can be quite tedious if there are many rows to select so named ranges are a good option (you can also use keyboard shortcuts).

Your result is a rectangular block (Figure 5.2). You can easily make a formula to calculate the correlation between the first two columns but copying the formula to the rest of your correlation matrix is awkward. You can fix cell references using the $ but this is only viable for copying a correlation formula across rows. If you try to copy down a column you can fix the row references but you cannot get Excel to use the next column along in the formula.

With a bit of fiddling you can make this work:

1. Make a framework for your correlation matrix; you need the variable headings across the top and down the side (see Figure 5.2).
2. The top-left entry is a correlation between a variable and itself, this is always 1. Either enter 1 or skip to the cell below.
3. You now want to correlate the first column with the second. Use CORREL to carry this out.
4. Edit the formula you made in step 3. You want to fix all the row references by preceding them with $ symbols.
5. Copy the formula to the cell beneath. The result will be identical. This row will need two entries ultimately. Each subsequent row will need an extra entry.
6. Now edit the formula in step 5. You want to alter the column references of the second sample; this will most likely be the next column along. Now this row of the matrix will contain two correlations. Copy the formula to the right; the result will be identical.
7. Click in the result in step 6 and then in the formula bar. You'll see the cell selection borders. Drag the border on the left to the column adjacent (to the right). The appropriate columns should now be selected and the result will update; press Enter to return to the worksheet.
8. Now click in the cell to the left, you need to copy this down one more (assuming there is another row in the matrix). The result will be identical of course.
9. You could repeat steps 6–7 but there is another way. Alter the second sample reference for the appropriate column (use the keyboard or drag the selection outline). Then fix the column reference of the second sample using $ symbols.
10. Now when you copy this formula to the right, only one of the columns will alter.
11. Repeat as necessary to complete more rows of the matrix. Each time you copy an entry down to the next row you'll need to alter the column reference and fix it with $ symbols.

If this sounds tedious, it is! Using named ranges can take some of the tedium out of the process but you'll still have to manually edit each cell of your correlation matrix. However, the editing required is easy to keep track of. In the following exercise you can have a go at using named ranges to make a correlation matrix.

Have a Go: Use named ranges to make a correlation matrix

You'll need the spreadsheet *NYenvironmental.xlsx* for this exercise. The data shows ozone measurements from New York in 1973 (May to September). There are also columns for solar radiation, temperature and wind speed.

↘ Go to the website for support material.

1. Open the *NYenvironmental.xlsx* spreadsheet; there is only one worksheet, *Data*. The first column contains a simple reference number containing the observation. The other columns are *Ozone* (the response variable) and others representing three predictor variables, *Radiation*, *Temp* and *Wind*.
2. Click once anywhere in the block of data. Then type Ctrl+A on the keyboard; this should select all the data.

3. Now click the *Formulas* > *Create From Selection* button (it is in the *Defined Names* section). From the dialog box that opens choose the option *Top row*, and make sure all other boxes are unticked. This will take the names of the cell ranges from the first row of the columns. Click *OK* to complete the operation and return to the worksheet.

4. Click once anywhere in the worksheet to deselect the data. You can view the named ranges using the *Name Manager* button. Click *Close* once you've finished looking at the named ranges.

5. Make a new worksheet to hold the resulting correlation matrix using the *Home* > *Insert* > *Insert Sheet* button.

6. Return to the main data and select the cells B1:E1, which contain the variable names. Copy to the clipboard, and then navigate to the new worksheet. Click in cell B1 and paste. Then click in cell A2 and use *Paste Special* (*Home* > *Paste* > *Paste Special*). Tick the *Transpose* box (near the bottom right) and then *OK* to complete the operation. You should have repeated the column headings but in a column, forming the left side of the correlation matrix.

7. Click in cell B2 and enter a 1 (number one), which represents the correlation between *Ozone* and itself. Repeat this operation along the diagonal of the correlation matrix.

8. Click in cell B3 and type a formula to calculate the correlation between *Ozone* and *Radiation*. Use the named ranges as in: =CORREL(Ozone, Radiation). Note that the formula is not case sensitive so you can use all lower case if you like.

9. Now copy the cell containing the formula to the clipboard and paste to all the remaining cells. They will be identical because named ranges are fixed.

10. Click in cell B4 to select it, then click the formula bar and edit the formula. You want to alter the *Radiation* named range to *Temp*. Press Enter to set the formula.

11. Click in cell B5 and change the formula to incorporate the correct named ranges. Repeat this process for the remaining cells. You can see which variables should be correlated from the headings of the correlation matrix. It is easy to alter the names to match. Your completed matrix should resemble Figure 5.3.

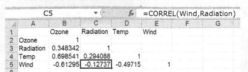

Figure 5.3 A correlation matrix created using named ranges.

You can edit the table as you like, altering the number format for example. There is no need to complete the top part of the matrix, as this is simply a repeat of the "bottom triangle".

Using named ranges is slightly tedious but it is preferable to messing around with $ and the mouse and cell references. If you have many columns then the *Analysis ToolPak* is a lot easier but you have the limitation of having to get all or nothing (as you need contiguous data).

If your data contains grouping (factor) variables there is no way you can get the *Analysis ToolPak* to work for you, and you'll have to use the CORREL function. You'll see how to do this in the next section, which also builds on the preceding exercise (so it is worth following).

5.1.2 Correlation matrices using grouping variables

When you have grouping variables you can still get correlation matrices but you have to work a bit harder. The *Analysis ToolPak* is no good for this work and you'll need the CORREL function. You could use *Sort* tools to arrange your data in order and then select the appropriate rows to use but this has lots of limitations. You could also use *Advanced Filter* operations to create subsets of your data (see Section 2.2.3). This is also quite long-winded.

The key to success is a pivot table and index variable(s). You need to be able to get a subset of your data for each level of the grouping variable; the *Report Filter* section of the pivot table is the way to do this. The index variable is the way you get the data displayed as a sample (that is the individual observations) rather than a summary. You've seen pivot tables used in helping to check for errors (Section 3.1.2); the general process goes like this:

1. Make an index for the grouping variable you want to use. The COUNTIF function is the way to do this (see Section 2.1.1).

2. Now make a pivot table; use the *Insert >PivotTable* button and place the result in a new worksheet.

3. Drag the index variable from the *Field List* at the top to the *Row Labels* box at the bottom. This will form the replicates (that is, the individual observations).

4. Drag the grouping variable to the *Report Filter* box. The default is to combine all the levels of the grouping variable, but you'll use a filter later.

5. Now drag the variables you want to correlate to the *Values* section. Each field you drag will form a separate column labelled *Sum of X*, where *X* is the name of the variable.

6. Once the pivot table is made go to the *PivotTable Tools > Design* menu and turn off any totals using the *Grand Totals* button.

7. The top of the worksheet should show the level of the grouping variable currently being displayed as *(All)*, usually in cell B1. Since all levels are being shown the number of rows displayed is the maximum number of observations in any group.

8. Use the mouse to select all the data in the pivot table, include the heading names but not the *Row Labels*. You are going to make named ranges for these data.

9. Use the *Formulas > Name Manager* button and delete any existing entries (they are easy to make anew). Click *Close* to finish.

10. Click the *Formulas > Create From Selection* button to open a new dialog box. Make sure that the only box ticked is *Top Row*. Then click *OK* to construct the named ranges.

11. Click once anywhere in the worksheet to deselect the cells. Now click the *Name Manager* button to open the *Name Manager* dialog box (see Figure 3.39).

12. At the moment the names are a bit clumsy, *Sum_of_X* and so on. Double-click the top entry to open a new window where you can enter a new name. Enter a sensible name. You should have removed any existing named ranges in step 9. Press Enter or click *OK* to set the new name.

13. Repeat step 12 for the other named ranges. Note that you can use the arrow keys to move down the list and the Enter key to open an item (quicker than using the mouse!). Click *Close* when you've edited all the names.

14. You can now make a correlation matrix using the named ranges you just made. You can place this in a new worksheet or somewhere near the pivot table.

15. Now you can use the *Report Filter* to view the correlation matrix for any single level of the grouping variable. There should be a dropdown icon in cell B1; see step 7. Click the dropdown icon and select a group from the filter dialog box. Click *OK* when you are done and the pivot table will update to show the samples for that level of the grouping variable. The correlation matrix you made will remain in place but it will be updated to show the correlation coefficients for this subset of the data (Figure 5.4).

	B22			f_x	=CORREL(Al,Mg)	
	A	B	C	D	E	F
1	Site	IsleThorns				
2						
3	Row Labels	Sum of Al	Sum of Fe	Sum of Mg	Sum of Ca	Sum of Na
4	1	18.3	1.28	0.67	0.03	0.03
5	2	15.8	2.39	0.63	0.01	0.04
6	3	18	1.5	0.67	0.01	0.06
7	4	18	1.88	0.68	0.01	0.04
8	5	20.8	1.51	0.72	0.07	0.1
9						
18						
19		Sum of Al	Sum of Fe	Sum of Mg	Sum of Ca	Sum of Na
20	Sum of Al	1				
21	Sum of Fe	-0.70631622	1			
22	Sum of Mg	0.98458031	-0.63501	1		
23	Sum of Ca	0.8618374	-0.45652	0.8005745	1	
24	Sum of Na	0.76839507	-0.24721	0.7586527	0.78266	1

Figure 5.4 A correlation matrix constructed from a pivot table and named ranges. If the pivot table is updated (using the *Report Filter*) the correlation matrix updates too.

You can quickly alter the group displayed using the *Report Filter*; note that you can do this from the worksheet (usually cell B1) or the *PivotTable Field List* (click the grouping field in the list at the top and set the *Filter* as required).

If you have more than one grouping variable you can switch between them easily using the *PivotTable Field List* task pane. You can also have more than one grouping variable, in which case you'll require an appropriate index variable too.

In the following exercise you can have a go at using named ranges and a pivot table to make a correlation matrix for yourself.

Have a Go: Link a correlation matrix to a pivot table

You'll need the *Pottery.xlsx* file for this exercise. The data show chemical composition of various pottery sherds from several sites. There are five columns relating to the chemicals and one for the site name (the grouping variable).

↘ Go to the website for support material.

1. Open the spreadsheet, the data are in the *Data* worksheet. Column H contains an index, *Index*, which is based on the *Site* variable. You could easily make a correlation matrix using the entire dataset but it would be good to see if the different sites have different correlations.

2. Start by making a pivot table, use the *Insert >PivotTable* button and place the result in a new worksheet. The *Site* variable is the grouping to use so drag this field from the list at the top to the *Report Filter* section.

3. Now drag the *Index* field item to the *Row Labels* section; this forms the observations.

4. There are five items representing different chemicals, drag each of the fields to the *Values* section, you might as well drag them in the order they appear, *Al*, *Fe*, *Mg*, *Ca* and *Na*. You should see a new column appear in the pivot table each time you add a field.

5. Click the *PivotTable Tools > Design* menu on the ribbon and use the *Grand Totals* button to remove totals (*Off for Rows and Columns*).

6. Click the *Formulas > Name Manager* button. If you see previously entered named ranges delete them (highlight the entries and use the *Delete* button). Click *Close* to continue.

7. Use the mouse to highlight all the composition data in the pivot table, that is cells B3:F17. You ought to have the five variable labels and all their data selected. Now click the *Formulas > Create from Selection* button and choose *Top Row* as the only option. Click *OK* and the names are set.

8. Click once in the pivot table to deselect the data. Now click the *Name Manager* button. Double-click the first entry and alter the name to Al. Press Enter to set the new name.

9. Use the down arrow to move to the next named range. Press Enter to open this entry for editing. Alter the name to Ca and press Enter. Repeat the process for the other named ranges so that they are named Fe, Mg and Na. Then click *Close* to return to the worksheet.

10. Use the mouse to select the headings of the table; you only want the variables you're going to correlate so select cells B3:F3. Copy them to the clipboard. Click in cell B19 and paste the cells to form the top of the correlation matrix.

11. Now click in cell A20 and then use the *Home > Paste > Paste Special* button. You can also right-click and select *Paste Special*. Make sure the *Transpose* option is ticked and then select *OK*. You should now have pasted the row headings for the correlation matrix.

12. Click in cell B20 and enter a 1 (number one) to represent the correlation between *Al* and itself. Fill in the other cells on the diagonal (C21, D22 and so on).
13. Now click in cell B21 and type a formula to calculate the correlation between *Al* and *Fe*, `=CORREL(Al, Fe)`. Fill out the rest of the correlation matrix in a similar manner.
14. To obtain a correlation matrix for each grouping you need to click the drop-down icon in cell B1 and select a sample from those available. The pivot table will alter to show the appropriate data. The correlation matrix will update to show the correlations for the sample you selected.

If you click in the pivot table to bring up the *PivotTable Field List* task pane, you can activate the *Report Filter* by clicking on the *Site* field name in the list in the top box. You'll need to click the triangle dropdown icon on the right that appears when you select the field.

In the preceding exercise you placed the correlation matrix underneath the main pivot table. You can put the matrix in any worksheet but it is a good idea to keep it in the same worksheet as the pivot table to remind yourself that the items are linked.

Tip: Freezing panes
If your pivot tables have a lot of rows you will not be able to see the *Report Filter* and the correlation matrix at the same time. You can use the *View > Freeze Panes* button to freeze a portion of the window, allowing you to keep the correlation matrix in view and still scroll to the top of the pivot table.

5.1.3 Highlighting correlations

A correlation matrix makes a good summary of the relationships between the variables in your dataset. However, it is still a basic table of numbers and if you have a lot of variables it would be useful to be able to pick out the more important relationships.

A correlation coefficient varies from a value of -1 to +1, with zero in the middle. A positive value shows you that as one variable increases in magnitude, so does the other. A negative correlation shows that as one variable increases, the other decreases in magnitude. Smaller values that are nearer to 0 than to 1 show a weak relationship, whilst values nearer to 1 show a strong relationship.

Assessing the strength of a correlation
So, how do you know if a relationship is a strong one or a weak one, and is the relationship statistically important? For the Pearson's product moment, which is used in the CORREL function and the *Analysis ToolPak*, the statistical importance is related to

the number of observations. This is called the *degrees of freedom*, which is related to the sample sizes as follows:

```
df = # observations -2
```

Since you must always have the same number of observations in each sample you subtract 2 from the number of pairs of observations. Once you have the degrees of freedom you can look up the critical value in a table of statistical values (Table 5.1).

Table 5.1 Correlation coefficients and level of significance (for Pearson's product moment). The correlation coefficient is statistically significant if equal to or greater in magnitude than the tabulated value for the appropriate degrees of freedom (df) and level of significance.

df	5%	1%	df	5%	1%	df	5%	1%
1	0.997	1.000	16	0.468	0.590	60	0.250	0.325
2	0.950	0.990	18	0.444	0.561	70	0.232	0.302
3	0.878	0.959	20	0.423	0.537	80	0.217	0.283
4	0.811	0.917	22	0.404	0.515	90	0.205	0.267
5	0.754	0.874	24	0.388	0.496	100	0.195	0.254
6	0.707	0.834	26	0.374	0.478	125	0.174	0.228
7	0.666	0.798	28	0.361	0.463	150	0.159	0.208
8	0.632	0.765	30	0.349	0.449	200	0.138	0.181
9	0.602	0.735	35	0.325	0.418	300	0.113	0.148
10	0.576	0.708	40	0.304	0.393	400	0.098	0.128
12	0.532	0.661	45	0.288	0.372	500	0.088	0.115
14	0.497	0.623	50	0.273	0.354	1000	0.062	0.081

In Table 5.1 you see a column for degrees of freedom (df) and corresponding critical values labelled 5% and 1%. If the value you obtained is equal or greater than the critical value (ignore the sign if the correlation is negative), for the appropriate degrees of freedom, then the relationship is said to be statistically significant.

Essentially this table gives you the maximum values that could be obtained simply by random chance. The 5% means that values in that column could be obtained 5% of the time, when you use random values. So, if your value is the same, or larger, there is only a 5% chance that you'd get this result down to random chance. Put another way, you can be 95% sure that your correlation is not simply down to random chance. The 1% column gives larger values; it represents the 1% situation so it is harder to get a random correlation. Usually the 5% value is used as a set cut-off point but some branches of science require more stringent levels. Notice that the table starts with df = 1; this would be the situation with three observations (df = obs–2), which is the minimum number you can sensibly use to compute a correlation.

Just because your correlation is statistically significant does not necessarily mean that one variable **causes** the other to alter in some way. There is a link between the variables but it may not be direct. Keep the following mantra in mind: "correlation does not imply causation".

So, you can look through your matrix of correlations and look for values that are greater than a particular value. This is not especially easy, particularly if you have a large correlation matrix. It would be nice if you could highlight the important values automatically. You'll see how you can do this next.

Conditional formatting

You can get Excel to format cells according to their contents; this is called *conditional formatting*. In a correlation matrix you want to see two things highlighted:

- Positive or negative values.
- Values at or above the critical threshold.

The conditional formats are controlled using the *Home > Conditional Formatting* button, which has a dropdown that contains a range of options (Figure 5.5).

Figure 5.5 The *Home > Conditional Formatting* button contains options for formatting cells based on their content.

At first glance the options are rather bewildering but you can make life a little easier by using the *Manage Rules* option, which renders the others superfluous!

You can set formatting rules for the entire worksheet or a range of cells. You can set the background cell colour or the font and style of the contents. In fact you can set more or less any cell formatting. The formatting is easy to set up and modify:

1. Use the mouse to highlight the cells that you want to apply conditional formatting to. This will probably be your correlation matrix.
2. Click the *Home > Conditional Formatting > Manage Rules* button. This opens the *Conditional Formatting Rules Manager* dialog box.

3. The top field should be set to *Current Selection*, but if not alter it. Now create a new rule using the *New Rule* button, which opens the *New Formatting Rule* dialog box. This contains several options; select the *Format only cells that contain* option (Figure 5.6).

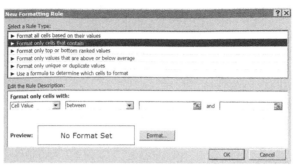

Figure 5.6 The *New Formatting Rule* dialog box allows you to set conditional formats based on cell values.

4. In the *Edit the Rule Description* section, alter the second box to read *greater than*, using the dropdown. Then in the last box type a number above which the formatting will apply. You can also point to a cell value in the spreadsheet. Generally you'll look up in the critical values table the appropriate value to use.

5. Now click the *Format* button to open the *Format Cells* dialog box. Set the formatting how you like (the simplest is to use bold font) and then click *OK* to return to the *New Formatting Rule* dialog box.

6. Click *OK* from the *New Formatting Rule* dialog box to apply the format and return to the *Conditional Formatting Rules Manager* dialog box.

7. You can delete or edit the new rule from the *Conditional Formatting Rules Manager* dialog box. At this point you want to set a new rule for negative correlations so click the *New Rule* button to open the *New Formatting Rule* dialog box. This contains several options; select the *Format only cells that contain* option again (Figure 5.6).

8. Now select *less than* as the option and an appropriate value below which the format will apply. Click the *Format* button and format as you like but different to before, red font color is helpful (Figure 5.7).

9. Click *OK* several times (three in total) to apply the formatting and return to the worksheet. The formats should now be applied to the cells you highlighted.

You can use *greater than or equal to* or *less than or equal to* as the criteria when formatting since the critical values are generally determined in this manner (Table 5.1).

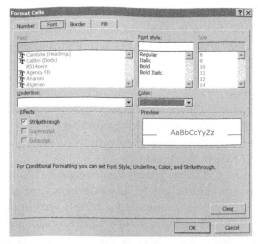

Figure 5.7 The *Format Cells* dialog box opens during conditional formatting of cells, allowing you to select a range of formatting options.

Tip: Highlighting non-rectangular cell ranges

If you use the mouse to select cells the default is to select rectangular blocks. If you wish to select a triangular range, such as in your correlation matrix, you can use the Ctrl key to select cells that are not adjacent.

It is useful to be able to highlight the important relationships in your data. Try to keep the conditional formatting simple; bold font for significant positive correlations and red for significant negative works quite adequately. In the following exercise you can have a go at highlighting some correlations.

Have a Go: Conditional formatting in a correlation matrix

You'll need the spreadsheet *NYenvironmental.xlsx* for this exercise. The data shows ozone measurements from New York in 1973 (May to September).

↘ Go to the website for support material.

1. Open the *NYenvironmental.xlsx* spreadsheet; there is only one worksheet, *Data*. The first column contains a simple reference number containing the observation. The other columns are *Ozone* (the response variable) and others representing three predictor variables, *Radiation, Temp* and *Wind*.

2. Use the *Data > Data Analysis* button to open the *Analysis ToolPak* and then select the *Correlation* option. Click *OK*, then select the data in the *Input Range* box. Tick the box that says *Labels in first row* and choose to put the results in cell G2. Click *OK* to complete the operation and create the correlation matrix.

3. Highlight the values for the correlations in the correlation matrix. Use the mouse to select the first values in the *Ozone* column, then hold the Ctrl key and select the values in the *Radiation* column; keep the Ctrl key pressed and click the final cell in the *Temp* column. You've now highlighted cells in a triangular portion.

4. Now click the *Home > Conditional Formatting > Manage Rules* button to open the *Conditional Formatting Rules Manager* dialog box.

5. Click *New Rule* then select the *Format only cells that contain* option from the *New Formatting Rule* dialog box.

6. Set the options to: *Cell value, greater than or equal to*. In the last box type the value 0.2, which is about right for the number of observations (see Table 5.1). Now click the *Format* button and set the formatting to bold font. Click *OK* to return then *OK* again to return to the *Conditional Formatting Rules Manager* dialog box.

7. Repeat step 5 and set up a new rule. This time use *less than or equal* to as the option and type the value -0.2 in the final box. Set the format to red font colour.

8. Now click *OK* three times to set the formatting and return to the worksheet. Your correlation matrix should now be formatted (Figure 5.8).

G	H	I	J	K
	Ozone	Radiation	Temp	Wind
Ozone	1			
Radiation	0.348342	1		
Temp	0.698541	0.294088	1	
Wind	-0.61295	-0.12737	-0.49715	1

Figure 5.8 Conditional formatting can help you spot important relationships in your correlation matrix.

You can see that some cells are bold and others are red, corresponding to the statistically significant positive and negative correlations respectively. You can see fairly easily that ozone is linked to the predictor variables, radiation and temperature are linked in a positive manner and wind is linked in a negative way. You can also see that some of the predictor variables are correlated to one another.

Highlighting important relationships using a correlation matrix and conditional formatting is useful and worthwhile but for a more detailed overview you ought to use visual methods; you'll see some options next.

5.2 VISUALIZING THE CORRELATIONS

A correlation coefficient is a useful summary of the relationship between two variables but for a more complete picture you need a graphical summary. Graphs also make it easier for others to see the relationships in your dataset. You can also use graphical methods to help you determine if your assumptions of normality are met. If they are not then you may have to consider alternative methods of analysis. You'll see how to check the assumptions later (Section 5.2.3) but before that you'll look at ways to visualize the patterns in your dataset.

5.2.1 Sparklines

As you saw in Section 4.3.2, sparklines are mini graphs that give values for a single range of cells and are presented within the confines of a single spreadsheet cell. They are available from the *Sparklines* section of the *Insert* menu on the ribbon. They are not really good for looking at correlations because you cannot get a scatter plot type of sparkline. The nearest option is the *Line* style of sparkline.

However, because they are quick to insert and small they have some usefulness and can be pressed into service. They work best if the intervals between the measurements are equal, which usually happens when you have time-sensitive data. You'll see more about sparklines later (Section 6.2.1), when you look at exploring time-related data (Chapter 6). To use sparklines follow these steps:

1. Click the *Line* button in the *Sparklines* section of the *Insert* menu.
2. Select the cells that form the data you want to chart; note that you should not include the heading; only include the numeric values.
3. Choose a place to put the sparklines; the number of cells you highlight should equal the number of columns of data you selected. Now click *OK* to complete the operation and insert the sparklines.
4. You should now see your sparklines but each one relates solely to one column. Click once in the block of data then use the *Data > Sort* button and rearrange the data in order of one of the variables; the response variable makes the best choice.
5. The sparklines occupy a single cell each so make the appropriate row taller by dragging the border in the row heading section. This makes the patterns easier to see (Figure 5.9).

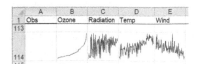

Figure 5.9 Sparklines are not the best choice to display correlations but if the response variable is sorted you can gain some idea of the patterns of relationships between variables.

You do have to be careful when using sparklines in this way, because your data are simply presented sequentially and with an equal interval. However, they are a quick way to look at some of the patterns in your data.

Tip: Filters and sparklines
If you use sparklines and then apply a filter, the chart will alter to reflect the filtered data. This can be useful if you have data with a grouping variable.

5.2.2   Scatter plots

Scatter plots are the ideal way to visualize the relationship between two variables. Once you have made a scatter plot Excel can show the link between the variables right there on the chart. You can use the *Chart Tools > Layout > Trendline* button (Figure 5.10) to add a line of best fit, as well as the mathematical formula that links the variables and a measure of how good the correlation fit is.

Figure 5.10 The Trendline button on the *Chart Tools > Layout* dialog allows you to place lines of best fit and other options.

You saw how to make scatter plots earlier, when you looked at checking your data for errors (Section 3.3.2). You can use exactly the same methods to produce scatter plots of the various combinations in your data. This is where the correlation matrix you made earlier comes in useful; you don't have to chart every combination, just the more interesting ones.

The simplest way to explore the correlations graphically is to build a scatter plot using one pair of variables and then use the data selection borders to move the charted data. If you add a *Trendline* you can see the general form of the relationship and the correlation coefficient too.

Note: Curvilinear trendlines
Trendlines (lines of best fit) do not have to be straight. Other mathematical relationships are possible. Excel can add logarithmic and polynomial lines of best fit for example, which produce curved lines.

The correlation coefficient that you can display is based on the coefficient that you calculate using CORREL. However, the value is shown as the Pearson coefficient squared; this is usually called the R^2 value. The R-squared value is of course always positive (because a negative value squared becomes positive) so you cannot relate directly to the table of critical values (Table 5.1) unless you take the square root (or square the table).

The dynamic selection outlines only appear if your chart is in the same worksheet as the data. This is slightly inconvenient, especially with large datasets, but there is a way to make the process easier:

1. Click once in the main block of data, then press Ctrl+A on the keyboard to highlight all the data.
2. Now click the *Formulas > Create From Selection* button and make named ranges for the variables. Take the names from the *Top row*.
3. Click once in the data to deselect everything. Then click the *View > Freeze Panes* button. Select the *Freeze Top Row* option (Figure 5.11). This means that

the variable names will always be visible, even when you scroll down to the bottom of the dataset.

Figure 5.11 The *View > Freeze Panes* button allows you to keep the variable names visible even when you scroll to the bottom of the dataset.

4. Scroll to the bottom of the dataset and click in a cell a few rows beneath the last data row.
5. Click the *Insert > Scatter* button and choose a scatter plot with points only. A blank chart will appear. Use the mouse to move it to a convenient spot.
6. Now click the *Select Data* button, then use the *Add* button and select the data you want. Click the *Series name* box and then the cell that contains the *x*-axis data label, then use the named ranges to fill in the *x* and *y* data series.
7. Click *OK* a couple of times to complete the chart and fill in the data points.
8. Click once in the chart to make sure the *Chart Tools* menus are displayed. Then use the *Layout > Trendline* button. Select the *More Trendline Options* button to open the *Format Trendline* dialog box (Figure 5.12).

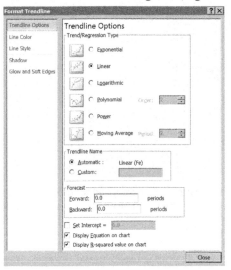

Figure 5.12 The *Format Trendline* dialog box allows access to all the options you need for trendlines.

9. Most of the time you will be looking for linear trendlines, so make sure this option is selected. Then tick the boxes to display the *Equation* and *R-squared* values. Click *OK* to complete the trendline.

10. Click once on the chart and then click and drag the Equation and R-squared values (they are in one textbox) to a sensible spot.
11. Format the chart as you require. Generally it is best to delete the legend and the gridlines. You might want to add axis titles to remind yourself which variables are being displayed. Use the *Chart Tools > Layout* menu for title options.
12. Click on any data point to activate the data selection borders in the worksheet. You can then drag the various items to new locations to redraw the chart. The Equation and R-squared values will update along with the data.

Tip: Dynamic selection and titles

If you set a *y*-axis title, don't type the name into the title box itself. Type = in the formula bar and then click on the cell containing the variable name. The main chart title will reflect the name of the *x*-variable and will update if you drag a selection outline.

If your data contains any grouping variables you can apply a filter to the data and update the display for any group (or combination). Click in the block of data and use the *Data > Filter* button. As long as your chart is below the main dataset it will remain visible (Figure 5.13).

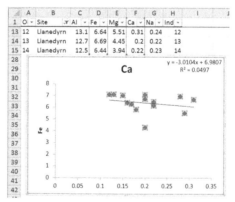

Figure 5.13 A scatter plot will update if a filter is applied to the main dataset. Clicking on any data point shows the data selection outlines, allowing you to alter the chart by dragging them to new variables.

You can make several charts but the dynamic selection only works if the chart is in the same worksheet as the data. If you want to make several charts you'll find it better to place each new chart in a separate worksheet. The named ranges enable you to construct a new chart with minimal fuss but of course if you have a lot of variables this could still take some time.

> **Tip: Multiple windows**
> Use the *View* menu to manage windows. The *New Window* button makes a new window and the *Arrange All* button allows you to arrange windows on screen.

If you do use multiple windows the R-squared values can be hard to read. Click on the chart to select the R-squared value then use the *Home* menu and make the font larger.

In the following exercise you can have a go at visualizing some correlations using a dataset containing a grouping variable.

Have a Go: Visualize correlation scatter plots with a grouping variable
You'll need the *Pottery.xlsx* file for this exercise. The data show chemical composition of various pottery sherds from several sites. There are five columns relating to the chemicals and one for the site name (the grouping variable).

↘ Go to the website for support material.

1. Open the spreadsheet; the data are in the *Data* worksheet. Column H contains an index, *Index*, which is based on the *Site* variable.
2. Click anywhere in the block of data, then press Ctrl+A on the keyboard to highlight all the data. Now click the *Formulas > Create from Selection* button and select the *Top row* option to set the names in the top row as the labels for the named ranges. Click the *OK* button to complete the operation then click once in the data to deselect.
3. Now click the *View > Freeze Panes* button and choose the *Freeze Top Row* option. You can now scroll to the bottom of the dataset and click in a cell just beyond the last row (cell B31 is ideal). You can press Ctrl+Down-arrow to go directly to the bottom of a column.
4. Click the *Insert > Scatter* button and choose a plain scatter plot (that is, one with points only). A blank chart is created; click and drag the chart so that it is positioned just below the last data row. You should still be able to see a few rows of the data above the chart, as well as the top row.
5. Now click the chart to activate the *Chart Tools* menus. Then click the *Design > Select Data* button to open the *Select Data Source* dialog box. Click the *Add* button to open the *Edit Series* dialog box.
6. Click in the *Series* name box and then click the cell D1, which contains the heading for the *Fe* variable. Now click in the *Series X values* box and type the named range for the *Fe* variable: `Data!Fe` (you can use lower case). Click in the *Series Y values* box, and delete any value in it and type the named range for the *Al* variable: `Data!Al`. Press Enter or click *OK* to return to the previous dialog. Click *OK* to close the dialog and return to the worksheet, where your chart is now formed.
7. The chart shows a legend, which you can delete, and a main title. Keep this as it shows the x-axis variable and will update if you alter the chart data. Click the gridlines and delete them.

8. Click in the chart to make sure the *Chart Tools* menus are active then use the *Layout > Trendline* button and select *More Trendline Options,* which opens a new dialog box. Tick the box labelled *Display R-squared value on chart.* The line type should be set to *Linear.* Click *Close* to return to the worksheet; you'll see the R-squared value on the chart but almost certainly in an inconvenient location.

9. Click on the chart then click and drag the R-squared value to the upper right corner of the chart. Now click the *Layout > Axis Titles* button and select a *Primary Vertical Axis Title.* Use the *Rotated Title* option; a title placeholder will appear on the chart.

10. Make sure the title of the *y*-axis is selected; click it once if you are not sure. Then click in the formula bar and type =. Now click the cell C1, which contains the label for the *Al* column variable. Now press Enter to link the *y*-axis title to that cell.

11. You are now showing a chart of *Al* against *Fe.* You can see the R-squared value of 0.6222 in the top right corner. Click on any data point once and you'll see all the points selected (sometimes it doesn't look like they are all selected but if it looks like the majority are it is fine).

12. You should see the selection outlines in the main data. Hover over the right edge of the border of the *Fe* column until the cursor looks like a four-headed arrow. Now click and drag the selection outline to the *Mg* column. The chart alters but not the title. Drag the outline for the cell D1 to cell E1; shifting from the *Fe* to the *Mg* column, the title should reflect the change. Notice that the R-squared value is also changed.

13. Now click out of the chart and in the block of data. Click the *Data > Filter* button.

14. Click the triangle icon in cell B1 to open the *Filter* dialog box for the *Site* variable. Filter the variable so that only the *Llanedyrn* site is displayed. Click *OK* to apply the filter. Look at how the chart changes with different sites. Drag the selection borders to alter the variables displayed. Remember that if you alter the *y*-axis variable you'll need to alter the axis title manually (as in step 10).

15. Click the chart to activate the *Chart Tools* menus, then click the *Design > Move Chart* button and select a new worksheet as the target for the chart. The default name is most likely *Sheet1* if you choose to place the chart as an object in the sheet, or *Chart1* if you have a dedicated sheet. Alter the name to reflect the displayed variables if you like.

16. Now click the *View > New Window* button to make a new window. Then click the *Arrange All* button. Select the *Vertical* option and also tick the box labelled *Windows of active workbook.*

17. You'll see your chart in both windows so click in the header of the left window to select it, and then click the *Data* tab at the bottom to display the data. Notice that the R-squared value is hard to read because it is quite small.

18. Click on the header of the right window to activate it, and then click the R-squared value on the chart to select it. Click the *Home* tab and make the font a larger size.

Try using the filter to alter the displayed group. If you wish to alter the variables you'll need to either make a new chart or use the *Select Data* button to alter the displayed variables (the selection outlines will not work).

If you make more charts you can simply use the *View* menu to make a new window and rearrange the windows; the *Tiled* option is good for multiple charts (Figure 5.14) but with more than about four windows on one screen things get a bit too crowded. You can always use the worksheet *Tabs* to switch windows. This is when having chart worksheets named sensibly is helpful.

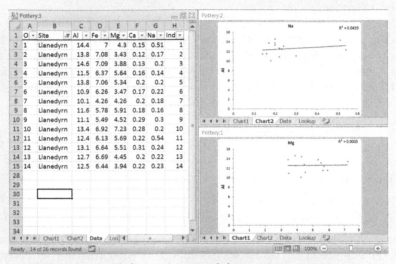

Figure 5.14 Tiled windows can help you view several charts at once.

If you have a filter active it will apply to all charts at the same time; you cannot have different charts showing different groups. If you want to compare groups you'll have to use the *Advanced Filter* and output the results to separate worksheets (see Section 2.2.3). Alternatively you can save your chart to a file, which effectively fixes the contents. You'll see more about this in Section 9.3.

Tip: Selecting chart elements
It can sometimes be tricky to click in just the right spot to select a chart element, especially if you are using multiple windows, as the charts can be small. To get around this click the chart to activate the *Chart Tools* menus. Then go to the *Layout* menu. You can select any chart element from the *Current Selection* section: the top item contains a dropdown so you can pick what you want. The same buttons are also available in the *Format* menu.

Excel is not really designed to let you display multiple charts all that easily or quickly; there simply aren't many shortcuts that you can use. Named ranges can make it easier to make a chart and the filter tools can let you explore groups but if you have a lot of variables to explore you'll need to set aside some time.

5.2.3 Checking assumptions of normality

Correlation and regression are similar kinds of analysis. In correlation you look at the strength (and direction) of the link between two variables. In regression you also examine the mathematical relationship between two (or more) variables. Earlier you saw how to add a trendline to a scatter plot. This trendline or line of best fit is added so that it fits neatly through the scattered points.

Slope, intercept and residuals

The trendline runs through the scattered points in such a way that the average distance between the points and the line is minimal. Some points may lie on the line but most are slightly above or slightly below. Usually there are about the same number of points above as below. The distances between the points and the trendline (measured along the y-axis) are called *residuals*. The best fit line is designed to minimize these residuals. The line can be described by a mathematical formula:

```
y = mx + c
```

This classic formula describes a straight line. The *m* is the slope and *c* is the intercept (where the line crosses the y-axis). You can calculate the slope using the SLOPE formula and the intercept using INTERCEPT:

```
SLOPE(y-variable, x-variable)
INTERCEPT(y-variable, x-variable)
```

You simply give the formulae the range of cells containing the response variable (the *y-variable*) and the predictor (the *x-variable*). It is important to get them the right way around!

There are other forms of regression but this straight-line linear regression is the most common. The regression relies on the properties of the normal distribution (the shape of the data, see Section 4.2.4). If your data are not normally distributed then you might have to carry out a different sort of analysis. So you should check that your data are normally distributed before you carry out any serious analysis.

It is usual to check the assumptions of normality on the results of linear regression as well as on the original data. Technically the linear regression is valid only if the residuals are normally distributed. This is not something you would necessarily do when you are exploring your data. Excel is not designed as a heavy-duty statistical package although it carries out regression analysis quite well and enables you to check the residuals quite easily.

Because the process of residual checking is relatively easy to demonstrate it is included in this guide to exploring regression data, as you'll see now.

Checking residuals

Once you have the slope and intercept you can use the $y = mx + c$ formula to describe the relationship between two variables (this is the equation added to a scatter plot when you add a trendline). The process can be extended to cope with more predictor variables; this is called *multiple regression* but here you'll see the simpler one-predictor case. Once you have the equation you can use it to predict the values in the *y*-variable (response) using values that you've already got from the *x*-variable (the predictor). You simply plug in each value of *x* to the formula using the slope and intercept values. The difference between these predicted values and the actual values form the residuals.

You can easily use the SLOPE and INTERCEPT formulae to work out the slope and intercept. Then you could determine the residuals for each value of the predictor variable. It is usual to subtract the predicted value from the observed value:

Residual = Observed – Predicted

Once you have these residuals you can make a histogram as you saw earlier (Section 4.3.1). You can use the FREQUENCY formula to work out the bins or use the *Analysis ToolPak* and let it work them out for you.

You can also use the *Analysis ToolPak* to calculate the residuals for you. One of the options in the *Analysis ToolPak* is *Regression*, which will not only carry out the regression but will give you the residuals and several potentially useful plots:

- *Line fit plot* – this is a regular scatter plot with the predicted values added; these are essentially the same as adding a trendline.
- *Residual plot* – this plots the residuals against the predictor variable.
- *Normal probability plot* – this is a way to visualize whether a sample is normally distributed. The response variable is plotted as a series of percentiles. If the plot looks straight, the sample is considered to be normally distributed.

The output does not produce a histogram or frequency plot of the residuals but it does produce the residual values, so it is easy to make a histogram from those. The way to proceed is as follows:

1. Open the spreadsheet containing the data; there is no need to select anything or to be in any particular worksheet but it is sensible to be able to see the main data.
2. Click the *Data > Data Analysis* button to open a dialog box, select the *Regression* option and click *OK*. This opens the *Regression* dialog box (Figure 5.15).

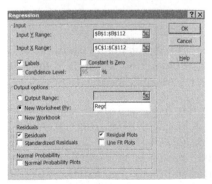

Figure 5.15 The *Regression* dialog box from the *Analysis ToolPak* allows you to calculate and visualize residuals.

3. Click in the *Input Y Range* box, then use the mouse to select the data (include the heading) for the response variable. You can avoid tedious dragging by using keyboard shortcuts: click the top cell (the one with the heading) then press Ctrl+Shift+Down-Arrow on the keyboard. The entire column of data is selected from the top row to the final entry in the column.

4. Now click in the *Input X Range* box. You will probably be near the bottom of the dataset now (see step 3), so click the bottom cell of the predictor variable, then press Ctrl+Shift+Up-Arrow on the keyboard to select all the way to the top.

5. Click the box entitled *Labels*, so that Excel knows that you've included the variable names.

6. In the *Output options* section select to place the result in a *New Worksheet Ply*; you can supply a name if you like.

7. In the *Residuals* section click the boxes entitled *Residuals* and *Residual Plots*.

8. You can select other options if you like but these are the basics to view the residuals. Once you are done, click *OK* to carry out the calculations and produce the charts in a new worksheet.

The new worksheet will contain some numerical output relating to the regression and any charts you decided to use. The chart for the residual plot will need some editing to make it more readable; you can use the *Chart Tools* menus to help you. The chart will have a main and axis titles that reflect the variable names you used in the analysis (Figure 5.16).

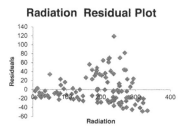

Radiation Residual Plot

Figure 5.16 A residual plot created by the *Analysis ToolPak*, showing the default settings (for Office 2010).

The residual plot (Figure 5.16) is potentially useful but it would be better to see if the residuals are normally distributed (this was indeed the purpose). You can now take the output from the regression performed by the *Analysis ToolPak* and make a histogram:

1. The output from the *Analysis ToolPak* shows several tables of results in two sections; the top is labelled SUMMARY OUTPUT and lower down is another labelled RESIDUAL OUTPUT. Scroll down to this lower output.

2. Look at the column headings; you will use the one labelled *Residuals*. Click the *Data > Data Analysis* button to open a dialog box, then select *Histogram* and click *OK*. This opens the *Histogram* dialog box (refer back to Figure 4.5).

3. Click in the *Input Range* box, where you'll select the data to be plotted (the residuals). Click the top cell (the top number not the column name) of the *Residuals* output, then press Ctrl+Shift+Down-Arrow to select all the residuals in the column to the bottom entry. This saves dragging the mouse down lots of rows.

4. In the *Output options* section, choose to place the results in a *New Worksheet Ply*; you can specify a name if you like.

5. Tick the box labelled *Chart Output*, which is right at the bottom. Click *OK* once you are done to make the frequency table and histogram in a new worksheet (Figure 5.17).

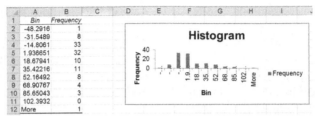

Figure 5.17 The result of the histogram routine of the *Analysis ToolPak* requires some editing to make it acceptable.

6. Edit the data and chart as required to make a sensible result (Figure 5.18). The data can be formatted to two decimal places and the chart bars need to be wider at the very least (see Section 4.3.1).

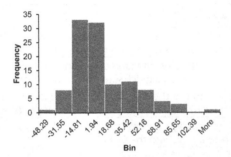

Figure 5.18 The histogram output from the *Analysis ToolPak* after editing. This histogram shows residuals from the regression between ozone and radiation in the *NYEnvironmental* dataset.

The checking of residuals is something that you will not undertake lightly, because it is rather time-consuming. Your early explorations of the data will help you to see the important relationships in your dataset and focus on the more important and relevant variables. As you work towards carrying out your statistical analyses you'll start to think more about details, such as the methods of analysis that you'll move on to next, and if the assumptions required by these methods are met by your data.

5.3 EXERCISES

1. The following correlation coefficients were calculated for pairs of variables, each with 20 observations. Which values are statistically significant?

   A) 0.43
   B) 0.51
   C) -0.43
   D) -0.51

2. You can use _____ to help you visualize the important relationships in a correlation matrix.

3. If you have a correlations or regressions dataset that includes grouping variables you can use a _____ to help you produce a correlation matrix.

4. Residuals are a way of showing how good the best-fit line is. TRUE or FALSE?

5. Look at the *women.xlsx* spreadsheet. The data represents the height (inches) and weight (pounds) of American women aged 30–39. The data were collected sometime before 1975. Calculate the residuals for the relationship; are they normally distributed?

↘ Go to the website for support material.

The answers to these exercises can be found in Appendix 1.

5.4 SUMMARY

Topic	Key points
Correlation and regression data	The link between variables. In correlation you have one response and one predictor variable. In regression you can have more than one predictor.
Correlation matrix	Shows the strength (and direction) of the links between pairs of variables. This gives you a quick summary of the relationships.
	You can use the CORREL function to determine the correlation between two variables. The *Analysis ToolPak* can produce a correlation matrix easily.
Named ranges	If you assign names to your variables you can use these to examine correlations with the CORREL function.
Correlations and grouping variables	If your data has grouping variables you can use a pivot table to rearrange your data, allowing you to explore correlations in those various sample groups.
Significant relationships	You can use a table of critical values to assess the statistical significance of a correlation.
Conditional formatting	Conditional formatting allows you to highlight the important relationships in a correlation matrix quickly and easily.
Sparklines	Sparklines can be used to make a line plot. Sparklines are not entirely suitable for correlation and regression data but can give an impression of the relationships between two variables, especially if the data are sorted in numeric order.

Topic	Key points
Scatter plots	The scatter plot is the most useful way to visualize the relationship between two variables. You can also add a line of best fit (trendline) to help assess the relationship. The straight-line equation and R^2 value (a measure of how good the fit is) can be displayed on your scatter plot.
	Using multiple windows can help you view several scatter plots at once.
Assumptions of normality	You can assess the slope and intercept of a line of best fit using the SLOPE and INTERCEPT functions.
	You can use the slope and intercept to help work out the residuals; that is the difference between observed and predicted values. These can be used in a histogram to check the assumptions of normality.
	You can use the *Analysis ToolPak* to compute residuals, which you can then examine with a histogram.

6

EXPLORING TIME-RELATED DATA

Sometimes you have data that have some important time element. How you approach the dataset depends somewhat on what you want to do with it. If you wish to follow one or more variables from one time period to another then you'll use the kinds of approach demonstrated in this chapter. If your data contains time-related information but you aren't necessarily going to follow from one interval to another then you either have regression data (see Chapter 5) or differences data (see Chapter 8). It is less likely that you have association data but it is possible (see Chapter 7). Each kind of dataset lends itself to a different way of being explored.

The importance of the time-related element is that you can follow measurements from one time period to the next; this leads to a particular way of visualizing your data, the line plot, which you'll see later (Section 6.2.2). These kinds of data are also particularly suitable for the sparkline type of chart, which you met earlier. You'll see sparklines again shortly (Section 6.2.1), but before that you'll look at how to summarize time-related data numerically.

6.1 SUMMARIZING TIME-RELATED DATA

It is difficult to summarize time-related data numerically! You can use basic summary statistics for the samples: averages, highs, lows and so on are all useful information. If you have grouping variables then you'll need to summarize group by group. Pivot tables are a useful exploratory tool. You'll see more specific information on using pivot tables on association data (Chapter 7) and differences data (Chapter 8).

If you have a simple sample you can create a numerical summary using various statistics; the QUARTILE function is especially useful as you can obtain the highest, lowest, middle and inter-quartiles quite easily. However, in most cases you want to be able to follow the course of a variable across the time periods under observation. The only sensible way to do this is with a pivot table.

6.1.1 Using pivot tables

For most time-related data a pivot table is essential. A pivot table allows you to rearrange your data in a sensible manner, summarizing the various samples according to the time variable. There are limits though; if you have a lot of separate times and few groups (or none), your pivot table will be the same as your original data. This is not helpful so you'll have to use some judgement as to when a pivot table is sensible. There are two main scenarios:

- You have one or more variables that you want to follow across a lot of time periods (observations).
- You have relatively few time intervals but your response variable has one or more grouping variables.

In the first case your pivot table will simply be too large to be sensible as a summary and you are better off using solely visual methods (although summaries of variables can be useful). In the second case a pivot table is the way to rearrange your data into a meaningful report.

If you have a lot of observations you may be able to make a grouping variable to split your data into smaller chunks. If you have daily observations for example it might be reasonable to group by month, week or some other interval. This allows you to view a summary of the overall data.

The general procedure in setting up a pivot table is as follows:

1. Make sure that your data has appropriate grouping variables. If you have lots of observations you might wish to split the data into months or some other interval (see Section 2.1.1 for notes about date variables).
2. Click once anywhere in the block of data. Then click the *Insert >PivotTable* button. The data should be selected automatically. Choose a *New Worksheet* as the location for the competed table, and then click *OK*.
3. You should now see the *PivotTable Field List* task pane. You need to drag the fields (column variables) from the box at the top to the appropriate boxes at the bottom to build the table. Start by dragging the variables that contain the data you want to follow into the *Values* box. These variables are generally response variables but could include predictor variables if they are numeric.
4. Now you need to drag the time variable to either the *Row Labels* box or the *Column Labels* box. If you use the former then the values are displayed in columns, with the dates down the side. If you choose the latter the values are displayed in a row.
5. You can drag grouping variables into one of the remaining free boxes. If you choose the *Report Filter* you'll combine all groups; you can then select one (or more) to display later. If you use the *Column Labels* or *Row Labels* box (the opposite of the one you used in step 4) your table displays multiple columns or rows.
6. You may need to alter the summary statistic displayed by the *Values* fields. The default is to use the sum so if the mean is more appropriate you need to click on a field in the *Values* box, select *Value Field Settings* and alter the summary statistic.

Exactly where you place the fields in the pivot table will depend on the result you require. You can drag fields about and rearrange the entire table quite easily until you get what you require. Because the arrangement is so dependent on the layout of your original data it is best to demonstrate using some specific examples. You can have a go at making some pivot tables in the following exercises.

Split a variable into smaller units

If you have a lot of observations, each with its own time period, it is most sensible to split the time into smaller chunks. How you do this will depend on how the data are arranged. If you have variables for the date you can make a new variable for the month-year combinations. You saw how to make new variables and deal with dates earlier (Section 2.1.1).

Making a new time variable creates chunks of time in your dataset and allows you to see the data summarizes. Each group in your new time variable will now contain several observations, whereas before there was probably only one per group (one per date). This means that when you use the pivot table you'll need to alter the summary statistic to *Average*, rather than the default *Sum*.

In the following exercise you can have a go at making a summary pivot table using a split-time grouping variable. This is a fairly simple exercise and will serve as a good foundation for more complicated pivot tables later on.

Have a Go: Use a grouped time variable to summarize large samples

You'll need the *airquality.xlsx* file for this exercise. The data are identical to those in the *NYEnvironmental.xlsx* file you've seen before, except that there are two additional columns, containing the month and day of the observations. The *Ozone* column shows ozone levels for New York in 1973. Other columns relate to solar radiation, temperature and wind speed.

↘ Go to the website for support material.

1. Open the spreadsheet; there is only one worksheet *Data*, which contains all the information you'll need.
2. Click once anywhere in the block of data, there is no need to highlight any data. Now click the *Insert >PivotTable* button. The data should be automatically selected so you should just select to place the result in a *New Worksheet* and click *OK*.
3. You should see the *PivotTable Field List* task pane; you'll need to drag the fields at the top to the appropriate boxes at the bottom to make the table. Start by dragging the *Month* field into the *Row Labels* box. You should see the pivot table beginning to form, the months (simple numeric labels) appear in a column as row labels.
4. Now drag the *Ozone* field into the *Values* box. Repeat the process for the *Radiation, Temp* and *Wind* fields. You should see four columns in the pivot table; they'll have titles such as *Sum of Ozone*.
5. The variables have been summarized by summing them, which is not entirely appropriate for these data. A mean would be better, since each *Month* contains several observations. Click on the *Ozone* field in the *Values* box, which opens a popup menu (Figure 6.1). Select the *Value Field Settings* option.

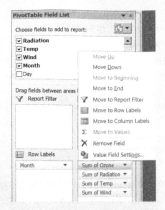

Figure 6.1 Click a field in the *Values* box to open a popup menu. The *Field Settings* option allows different summary statistics to be selected.

6. From the *Value Field Settings* dialog box select *Average* in the *Summarize value field by* box, and then click *OK*. The *Ozone* field will now update and display the mean instead of the sum.
7. Repeat step 6 for the other variables so that they all show the mean. The column names and the field labels will show *Average of Ozone* and so on.
8. The *Column Labels* box will display a field labelled ∑ *Values*. Try dragging this into the *Row Labels* section and swap the *Month* field into the *Column Labels* box. Notice what happens when both fields are in the same box.

You can move the *Month* and ∑ *Values* fields around to make a different looking table. Pick the arrangement that best conveys the information you want.

The pivot table is easily modified by moving the field items, and clicking them to alter the summary and formatting.

> **Note: Field order in pivot tables**
> You can have more than one field in any of the sections of a pivot table. The order of the fields is important. The topmost field forms the highest level of the hierarchy and lower fields are nested in groups below that. You can alter this hierarchy simply by dragging the fields into a new order.

> **Tip: Number formatting in pivot tables**
> You can set the number of decimal places in pivot table items. Click a field and select *Value Field Settings*. Then use the *Number Format* button.

There are other options that you can utilize, as you'll see in the following exercises.

Follow time-related data with one grouping variable

In the preceding exercise you had a lot of observations over time (daily records), but there were no experimental grouping variables. You made a grouping variable based on the month of observation in order to help summarize the data by reducing the number of time-periods to follow.

In other cases you'll have fewer time periods but will have other sorts of grouping variables. In the following exercise you'll see how to manage a simple situation where there is a single grouping variable.

Have a Go: Summarize time-related data with a single grouping variable

You'll need the *Butterfly and Year Records.xlsx* file for this exercise. The data show the number of various butterfly species (*Spp*) observed at a nature reserve in Scotland. The observations give the *Count*, which is repeated over several years. Note that only positive records are presented. If a species was not observed there is no record for it, in other words there are no zero entries. This is an important point.

↘ Go to the website for support material.

1. Open the spreadsheet; there is only one worksheet, *Data*. There are relatively few time periods (1996 to 2005) but you have a grouping variable, the species (*Spp*), of which there are 20. Notice that some of the *Count* data are non-integer. The counts were taken from several transects (lines along which butterflies were counted) and adjusted for area (lengths of the transects).

2. Click once anywhere in the block of data, then click the *Insert >PivotTable* button to start the process of making a pivot table. The data should be highlighted automatically (look for the marching ants) so choose to place the result in a *New Worksheet* and click *OK*.

3. Drag the *Count* field from the list at the top to the *Values* section at the bottom. Drag the *Year* field to the *Column Labels* section. Now drag the *Spp* field to the *Row Labels* section.

4. You'll see that the table now shows the various species in the column on the left and you can follow their abundance from year to year by reading across the columns. There are also grand totals for the rows and for the columns. Click on the table to activate the *PivotTable Tools* menus; these allow you to access many tools that allow you to manage your table (many tools can also be accessed by right-clicking a pivot table element).

5. Notice that there are gaps in the data; these relate to years when a particular species was not recorded. There are occasions when it is important to have a zero value, rather than a blank. You can turn blanks into zeroes (or any character) by accessing the PivotTable options. Click the *PivotTable Tools > Options >PivotTable* button, and then click *Options*. You should now see the

PivotTable Options dialog box (Figure 6.2). You can also access this dialog by right-clicking anywhere in the pivot table and selecting *PivotTable Options* from the popup menu.

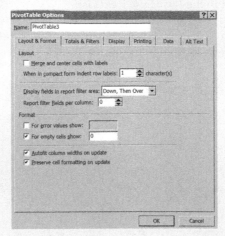

Figure 6.2 The *PivotTable Options* dialog box allows you to alter many settings for pivot tables.

6. The *PivotTable Options* dialog box has several tabs; you should see the *Layout & Format* tab. Click in the box that is labelled *For empty cells show* and type a 0 (zero) into the box. The tick-box should be enabled but if it isn't click to enable it. Click *OK* to close the dialog and you'll see the gaps replaced by zeroes.

7. You can filter the species using the dropdown icon on the *Row Labels* heading in the pivot table (Figure 6.3). You can also invoke the filter dialog from the *PivotTable Field List*; click on a field in the top section (you need to click on the right, over the triangle icon).

Figure 6.3 *Row Label* items can be filtered using a dropdown menu.

8. The filter allows you to view any combination of species in your pivot table. Try a few options. When you are done, recheck *(Select All)*. Notice that when a filter is operational you'll see a funnel icon in the header of the pivot table and by the field in the *PivotTable Field List*.

9. Try rearranging the *Year* and *Spp* fields in the *Row Label* and *Column Label* boxes. Try both fields in the same section, and in different order. End up with the *Spp* and *Year* fields in the *Row Labels* box and with *Spp* on top (Figure 6.4).

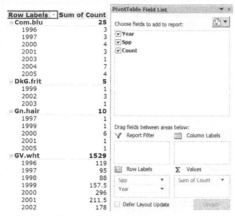

Figure 6.4 You can have multiple fields in one section, in which case the order is important. Note that blank entries do not display by default.

10. Look at the top entry (*Com.blu*) in the pivot table; it has missing years (1998, 1999, 2001). The same goes for all the species. So, even though you set blank entries to display zero they are still missing. You need to alter the settings for the *Year* field so that it will display missing entries. Click the *Year* field in the *PivotTable Field List* (in the *Row Labels* section). Select *Field Settings* to open the *Field Settings* dialog box. You can also right-click a year label in the pivot table.

11. Click on the *Layout & Print* tab and then tick the box labelled *Show items with no data* (Figure 6.5). Click *OK* when you're done to return to the pivot table.

12. Now you should see the entries that are blank showing 0. You can turn the zeroes off and on from the *PivotTable Options* (steps 5–6), so you could display gaps if you prefer.

13. Other options are available from the *PivotTable Tools* menus. Click on the pivot table and then the *Design* tab. Use the *Grand Totals* and *Subtotals* buttons to turn off all the totals.

14. The *Report Layout* button allows you to alter the layout of the table quite subtly. Try the various options (Figure 6.6). If you set *Show in Tabular Form* and *Repeat All Item Labels* you get a layout that is useful for copying and pasting for example.

Figure 6.5 You can alter the *Field Settings* so that entries with no data are shown in the pivot table.

Figure 6.6 The *Report Layout* button allows your pivot table to be laid out in subtly different ways.

15. Return to the *Compact Layout* using the *Report Layout* button and then drag the *Spp* field from the *Row Labels* section to the *Report Filter* section in the *PivotTable Field List* task pane.

Figure 6.7 The *Report Filter* is an overall filter for your pivot table. The name of the item displayed is shown at the top unless you display multiple items.

16. The *Report Filter* is an overall filter; to start with you'll see *(All)* displayed in the cell at the top of the pivot table (usually cell B1). You can alter the item that is displayed using the filter and the name will reflect the item selected. If you choose several items the name will display *(Multiple Items)* in cell B1 (Figure 6.7). You'll need to tick the box labelled *Select Multiple Items* to be able to select multiple items.

The *Report Filter* amalgamates multiple items via the filter and this might not be appropriate for your data. In this case you have counts, so it is appropriate. If you need to alter the summary statistic used you can simply click the field and alter the *Field Settings* (step 11) to show a more appropriate measure from the options available.

When you have multiple grouping variables things become more complicated but the same processes apply, as you'll see next.

Follow time-related data with multiple grouping variables

In the previous example you saw how to use a pivot table to arrange your data and follow a response variable across time periods. In that example there was only one grouping variable but you can easily extend your pivot tables to incorporate more. In the following exercise you can have a go at using a pivot table for a dataset that has several grouping variables.

Have a Go: Summarize time-related data containing multiple grouping variables
You'll need the *Butterflies and Site Management.xlsx* file for this exercise. The data show the abundance of four species of butterfly on a site in Southern England. The *Qty* column shows the abundance as an index calculated from counts and divided by the length of the transects. There are five different transects and these are subdivided into 12 sections. Some of the sections were managed in various ways and others not. The *Mng* column shows if there was active management or not. The *Year* variable shows the year the data were collected (1996–2007).

↘ Go to the website for support material.

1. Open the spreadsheet; there is only one worksheet, *Data*. You'll notice that there are zero entries for some records. You want to follow the abundances (*Qty*) across the time periods so start by making a pivot table. Click in the block of data then use the *Insert >PivotTable* button and place the result in a *New Worksheet*.

2. Drag the *Qty* field from the list at the top of the *PivotTable Field List* task pane to the *Values* section at the bottom; this is your response variable. The field will now read *Sum of Qty*, but you really need this to be an average. Click the field and then select *Value Field Settings* from the popup menu. Click the

Average option in the *Summarize value field by* section. Now click the *Number Format* button and alter the field to a *Number* format (the default 2 decimal places is fine). Click *OK* twice to return to the pivot table. The field should now display *Average of Qty*. It is the mean of course but Excel simply calls it *Average*; you cannot get a median in a pivot table.

3. Now drag the *Spp* field to the *Column Labels* section and the *Year* field to the *Row Labels* section. You've now got a table that averages the abundance of the species for all transects, sections and management. Notice that the pivot table has headings labelled *Row Labels* and *Column Labels* (Figure 6.8).

Average of Qty	Column Labels				
Row Labels	pbf	spbf	swf	wa	Grand Total
1996	0.99	0.47	7.30	0.65	2.35
1997	2.86	2.89	10.33	0.27	4.09
1998	0.82	1.89	2.77	0.30	1.44
1999	0.28	0.63	4.67	0.12	1.43
2000	0.54	0.62	2.89	0.24	1.07
2001	0.44	0.83	3.83	0.20	1.32
2002	2.26	1.99	4.61	0.26	2.28
2003	1.97	3.23	7.16	0.31	3.17
2004	3.45	5.59	10.09	0.88	5.00
2005	3.44	5.81	8.46	0.45	4.54
2006	3.12	3.53	21.17	1.49	7.33
2007	4.27	2.89	10.25	0.75	4.54
Grand Total	2.04	2.56	7.76	0.49	3.21

Figure 6.8 The *Row Labels* and *Column Labels* headings allow access to filter options but also make the columns wider. These headers can be switched on and off using the *Options > Field Headers* button.

4. Click in the pivot table to make sure the *PivotTable Tools* menus are active, then click the *Options > Field Headers* button. The table is a bit neater but you cannot see the filter dropdown icons. To access the filters you can click the fields in the top section of the *PivotTable Field List* (you'll need to use the triangle icon, which appears when you hover over with the mouse). Have a look at some of the filters but clear all filters before you move to the next step (there is a *Clear Filter* option).

5. Grand totals aren't especially relevant for these data so click on the pivot table and then use the *Design > Grand Totals* button to turn them off. Keep the subtotals in place for the time being (although none are displayed).

6. It would be interesting to compare the effects of management on the abundance, so drag the *Mng* field to the *Column Labels* section. Make sure you place the field underneath the *Spp* field. Now for each species you should see columns headed *no* and *yes*. You'll also see a subtotal for each item (species) which in this case is the mean. Each item (species) shows a - icon in the heading, if you click this icon it alters to a + and the levels underneath are collapsed to a single column (Figure 6.9). Try this out for some of the headings to see the effects.

Average of Qty	⊟ pbf	⊟ spbf		spbf Total
		no	yes	
1996	0.99	0.11	0.59	0.47
1997	2.86	0.15	3.84	2.89
1998	0.82	0.00	2.47	1.89
1999	0.28	0.24	0.75	0.63
2000	0.54	0.03	0.79	0.62
2001	0.44	0.13	1.04	0.83
2002	2.26	0.17	2.55	1.99
2003	1.97	0.06	4.20	3.23
2004	3.45	0.34	7.20	5.59
2005	3.44	0.07	7.56	5.81
2006	3.12	0.44	4.48	3.53
2007	4.27	0.43	3.74	2.89

Figure 6.9 The +/- icons allow you to collapse or expand the fields underneath the item you select. In this case the *yes* and *no* columns are collapsed to one column showing the mean.

7. Now drag the *Mng* field from the *Column Labels* section to the *Row Labels* section; place *Mng* underneath the *Year* field. You can experiment and drag the fields to different locations and explore the effects. Leave the *Report Filter* box empty for now. When you are finished restore the *Mng* field to the *Row Labels* section. Then click the *Design > Subtotals* button and turn all subtotals off.

8. You'll now explore the *Transect* field. You can start by dragging the *Transect* field into the *Row Labels* or *Column Labels* sections. The order of the fields in the sections has an effect on the table so try moving things around. You can also use the +/- buttons to collapse and expand columns and rows. Once you've had a go at this move the *Transect* field into the *Report Filter* section.

9. Your table should show the species in columns, split by the site management. At the top of the table you'll see *Transect* and *(All)* to remind you that you are using the *Transect* field as a *Report Filter* and that currently all the transects are selected. Use the filter to look at the various transects: you can click on the triangle icon by the *(All)* label or click on the field in the top of the *PivotTable Field List* task pane.

10. Now drag the *Mng* field to the *Report Filter* section. You can now apply two filters together; it does not matter which order they are in. Try applying a few filters and see the effects on the table.

11. You can also drag the *Section* field to the *Report Filter* section and use three filters! The *Section* variable is slightly odd because some sections have active management in one *Transect* but not in another. However, you can see that by rearranging the fields and using the *Report Filter* you can view the data in many different ways.

These data are grouped in a somewhat *ad hoc* manner! The different transects allow replicated data to be collected. Within each transect there are different sections and these have active management or they do not. You can get an overview by rearranging the pivot table in the following way: *Values: Average of Qty; Row Labels: Spp; Column Labels: Transect, Section, Mng (in that order).*

Things can get quite complicated when you have several grouping variables. However, the pivot table allows you to arrange and rearrange the various variables quickly and easily.

You used various filters in the preceding couple of exercises. These are useful in allowing you to view various chunks of your data. Later versions of Excel (from 2010 onwards) provide another way to access filters: the tool is called the *Slicer*, and you'll see how to use this next.

Pivot table Slicers

The *Slicer* tool is a way of using filters to manage a pivot table in a convenient and inter-active way. Once you have a pivot table you can create a slicer from the *PivotTable Tools* > *Options* menu. You can also use the *Insert* menu but you have to have clicked on the pivot table first, so you might as well use the *Options* menu!

The general process of making and using a slicer is as follows:

1. Make a pivot table. You do not have to incorporate all the fields into the table in order to use a slicer but the variables do have to be in the *PivotTable Field List*.
2. Click the pivot table to activate the *PivotTable Tools* menus; this opens the *Insert Slicers* dialog box, where you can choose the fields to use for the *Slicer* (Figure 6.10).

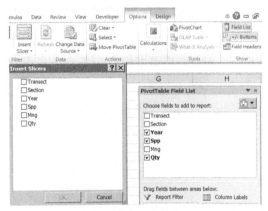

Figure 6.10 The *Insert Slicers* dialog allows you to select one or more fields from a pivot table to use as filters.

3. Use the check boxes to select the fields you want, and then click the *OK* button The filters appear as boxes in the worksheet (Figure 6.11). If you selected more than one field the boxes are cascaded.
4. When you click on a slicer window the *Slicer Tools* menu is activated; this gives you some editing options. The most useful ones are in the *Arrange* and *Buttons* sections (Figure 6.12).
5. Use the buttons in the *Arrange* section to reorder the windows; you can also simply drag the windows to convenient locations on screen.

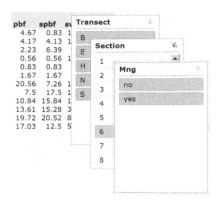

pbf	spbf	sv
4.67	0.83	1
4.17	4.13	1
2.23	6.39	
0.56	0.56	1
0.83	0.83	
1.67	1.67	
20.56	7.26	1
7.5	17.5	1
10.84	15.84	1
13.61	15.28	3
19.72	20.52	8
17.03	12.5	5

Figure 6.11 Multiple slicers appear cascaded on screen. You can move and edit each window independently.

Figure 6.12 The *Slicer Tools* menu includes tools that allow you to alter how the slicer windows appear.

6. Use the *Columns* button in the *Buttons* section to set how many field buttons are displayed in the selected slicer window. The default is for one button across; setting to a larger value usually makes it easier to work with the windows (Figure 6.13).

7. Use the mouse to resize and position the windows (Figure 6.13).

Figure 6.13 You can resize and arrange slicer windows with the mouse. Multiple columns can make the windows easier to manage.

8. Click the buttons in the slicer windows to create the filters. By default all items are selected; the first time you click a field it is selected and the others deselected (they will appear lighter, depending on your colour scheme).

9. Select multiple items using the Ctrl and Shift keys in conjunction with the mouse. This works in the standard Windows way: the Ctrl key selects (or deselects a single item), while the Shift key selects that item and everything between it and the previously selected item.

10. Clear the slicer from a window using the funnel with a red cross icon.
11. Delete the slicers you don't want by clicking on one and using the Delete key on the keyboard.

You can right-click on a slicer window to open a popup menu giving access to various options. The *Slicer* tool is useful, especially if you have a large and complicated dataset. It is a bit easier to manage your slicers than using multiple report filters (which is essentially what a slicer is).

A pivot table is useful because it allows you to rearrange your data easily, which you might want to do for a particular analysis. However, numerical summaries are useful but not as easy to follow as visual summaries, that is, graphs. You'll see how to visualize your time-related data in the next section.

Note: *Time Slicer* button

If you use Excel 2013 you can use the *Time Slicer* button. This looks for columns formatted with a date format (day, month and year), and allows you to view chunks of your data based on the date. If you've only got the year you'll need to fool the system by creating a new variable `DATE(yr_cell, 1, 1)`.

6.2   VISUALIZING TIME-RELATED DATA

A visual summary of your time-related data (or indeed any sort of data) is generally more accessible than a purely numerical one. The most usual type of graphical representation is a line plot. The line plot shows the magnitude of a variable at various intervals, the points being joined with a line. The line allows you to see how the variable alters over time.

There are two main types of plot you can produce:

- *Sparklines* – quick in-cell plots that give an overview of a single variable.
- *Line plots* – plots that show one variable on the *y*-axis with the *x*-axis being split into equally sized categories, which are usually time periods. You can add multiple variables to a line plot.

Both types are useful, although sparklines are less flexible, as you'll see next.

6.2.1   Sparklines

Sparklines are mini-charts that are created to fit inside a single cell of your spreadsheet. They are available from the *Sparklines* section of the *Insert* menu. You met sparklines earlier when looking at visualizing regression data (Section 5.2.1).

The general way of inserting sparklines is as follows:

1. Click the *Line* button in the *Sparklines* section of the *Insert* menu.
2. Select the data you wish to chart. Do not include any labels.

3. Select the range of cells to hold the sparklines, if you are charting columns you need to match the number of columns in the data. If you are charting by row, match the number of rows. The cells don't have to be adjacent to the actual data.
4. Click *OK* to insert the sparklines.
5. Click on a sparklines cell to activate the *Sparklines Tools* menu, there is only one item, the *Design* menu. You can use the tools in the *Design* menu to alter the appearance of the sparklines.
6. Alter the width and height of the cells as needed to make them display clearly. You can drag the row or column boundaries using the mouse or right-click to set a size explicitly (you can set multiple cells this way).

Sparklines are fairly quick to insert and there are various ways you can alter your sparklines to help you visualize your data, as you'll see shortly. You can create sparklines from pivot table data, which is especially helpful if you have grouping variables. This gives you a good deal of flexibility and allows you to carry out a visual exploration of your data with minimal fuss or effort.

There is little difference between using sparklines on raw data or on a pivot table; your approach depends on whether you have grouping variables or not.

Sparklines from raw data

Your data should always be set out so that each column is a variable and each row a single observation (Section 1.1.2). However, within this system you could have one of two scenarios:

- *Single dates* – each row is a single date and the columns are separate variables without grouping.
- *Repeated dates* – dates in rows are repeated because the columns contain grouping variables.

When you have single dates (that is, no two rows have the same date) the implication is that you do not have any grouping variables. If you do have a grouping variable then it will have only one level (in which case it is redundant). This situation is most likely when you have regression data: you'll have a response variable and one or more predictor variables, each with their own column. So, the rows represent the readings taken at a single time period. You've seen examples of this with the air quality data in the *NYenvironmental* and *airquality* datasets.

If you have repeated dates the implication is that you have grouping variables. Each row represents a single observation of the combinations of the groupings, so if you have several groups you'd have several observations for every time period. You've seen examples of this layout with the butterfly datasets. If you have a situation like this then you should make a pivot table first, and then add your sparklines.

Most times you want to follow the columns so your sparklines will be in a row; you'll chart how each column changes as you move down the rows, so you want your data sorted in date order. Often a convenient location is the bottom of the columns you are charting but it is possible (and often desirable) to place the sparklines in a new worksheet.

In the following exercise you can have a go at producing sparklines for data with single dates; this will give you some practice at the basics.

Have a Go: Insert sparklines for single-date data

You'll need the *airquality.xlsx* file for this exercise. These data are identical to the *NYEnvironmental.xlsx* file you've seen before, except that there are two additional columns, containing the month and day of the observations. The *Ozone* column shows ozone levels for New York in 1973. Other columns relate to solar radiation, temperature and wind speed.

↘ Go to the website for support material.

1. Open the spreadsheet; there is only one worksheet, *Data*, which contains all the information.
2. Make a new worksheet to hold the sparklines: click the *Home > Insert > Insert Sheet* button. You could place the sparklines at the bottom of the data columns but that can be inconvenient, especially when there are many rows of data.
3. Make sure that you are in the worksheet where the sparklines are to be placed. There is no need to highlight anything. Now click the *Insert > Line* button to open the *Create Sparklines* dialog box (Figure 6.14).

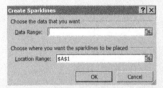

Figure 6.14 The *Create Sparklines* dialog box allows you to select the data and target location for sparklines.

4. Click in the *Data Range* box and then select the data. Click on the *Data* tab to return to the main data and start by clicking on cell B2. You can use the mouse to select the data or keyboard shortcuts: click in cell B2 and drag to select the cells in that row (B2:E2), then press Shift+Ctrl+Down Arrow on the keyboard. Now all the data you want are selected.
5. Now click in the *Location Range* box. You return to the new worksheet. You'll need to select the same number of cells in a row as there were columns of data (columns B:E). So, start by deleting anything already in the *Location Range* box then select cells B1:E1. Now click *OK* to make the sparklines.
6. Make the cells taller by dragging the border between rows 1 and 2 in the margin. The cursor alters when you are in the correct spot. You can also right-click in the border area for row 1 and select *Row Height* from the popup menu. Row height is given in pixels; choose a value of around 90.
7. Now alter the cell widths. You can drag the borders individually or select them all at once. Click in the column border area and drag to select the

columns B:E. Now right-click and select *Column Width* from the popup menu. Choose a value of around 15. Click once anywhere in the worksheet to deselect the cells.

8. Click once on any of the sparklines, this selects them all and activates the *Sparklines Tools* menu; click the *Design* tab (the only one in the *Sparklines Tools*) to show the options available (Figure 6.15).

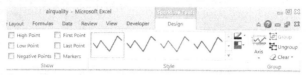

Figure 6.15 The *Sparklines Tools > Design* menu gives a range of tools for altering the appearance of sparklines.

9. In the *Show* section tick the boxes labelled *High Point* and *Low Point*. The appropriate data are shown in the sparklines but are probably hard to see. You can alter the colour of these points using the *Marker Color* button, which is towards the right in the *Style* section (Figure 6.16). Click the *Marker Color* button and alter the colour of the high and low points; set them both to red.

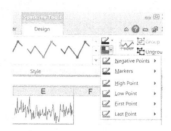

Figure 6.16 The *Marker Color* button allows you to alter the colours, so highlighting high and low points.

10. Each sparkline is scaled independently so that the lines fit most neatly into the cell. You can change the axis options using the *Axis* button (Figure 6.17). Alter the *Maximum* and *Minimum* to the *Same for All Sparklines* option.

11. Note how the sparklines all now show the same vertical scale but the patterns are harder to read. Alter the axes back to *Automatic for Each Sparkline*.

12. Each sparkline is contained in a cell and forms a background. This means that you can use the cells as regular cells. Add a label to the sparklines: click on cell B1 and type a formula to display the variable name; =Data!B1 will do nicely. You can also type = and select the cell with the mouse.

13. Now copy the formula across the row to fill the other sparklines. The names of the variables are displayed. Select all the sparkline cells and then alter the label appearance from the *Home* menu. You can alter the font colour and set the text to be centred vertically and horizontally (Figure 6.18).

Figure 6.17 The *Axis* button allows you to alter the scale for sparklines; the default is to scale each one independently.

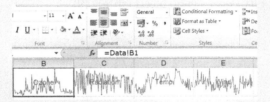

Figure 6.18 Since sparklines are in the cell background you can use the cells as regular worksheet cells. Here the cells point to the variable names.

Since the sparklines cells act like regular cells you can alter the background, borders and so on without affecting the actual sparklines.

In the preceding exercise you created the sparklines in a separate worksheet. Using a separate worksheet is not essential but it is good practice to keep your original data un-sullied. You can easily make a *New Window* (using the *View* menu) and arrange windows on screen.

When you have grouping variables you should make a pivot table first, which is what you'll see next.

Sparklines from pivot table data

When you have any grouping variables it is easier to make a pivot table first and then insert your sparklines from there. You can use the filters in your pivot table to view different chunks of your data. You saw how to use pivot tables earlier, when you summarized time-related data (Section 6.1.1).

Using sparklines in a pivot table is essentially the same as for raw data. You simply point to the cells in the pivot table that contain the data you need. If you apply any filters (for example using the *Report Filter*) the sparklines update. You'll almost certainly have to resize the columns containing the sparklines each time you update the table, but this is a minor issue.

If you have missing data, that is for some time periods your data are blank, rather than zero (0), your sparklines show gaps by default. You can alter the way missing data is handled in one of two ways:

- Change the *PivotTable Options* so that empty cells are shown as zero.
- Alter the settings for the sparklines.

In the first case you click on the pivot table then right-click and select *PivotTable Options* (or use the *PivotTable Tools > Options >PivotTable> Options* button). In the box labelled *For empty cells show* type a zero (0). Empty cells in the pivot table will now display as zero and the sparklines will update to show the gaps as 0, with all items joined by the line.

You can also alter the sparklines options directly:

1. Click on a sparkline to activate the *Sparklines Tools* menu.
2. Click the bottom of the *Design > Edit Data* button and select *Hidden and Empty Cells*. This opens a new dialog box (Figure 6.19).

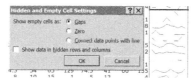

Figure 6.19 The *Hidden and Empty Cells* dialog (from the *Sparklines Tools > Design > Select Data* button) gives you options for dealing with missing values.

3. Select the *Zero* option and click *OK*.

If you alter the pivot table substantially you might end up overwriting the cells containing the sparklines; if so you can delete the sparklines easily using the *Clear* button on the *Sparklines Tools > Design* menu. The *Clear Selected Sparklines* option deletes the sparklines in the cells you selected. If you choose the *Clear Selected Sparkline Groups* you'll remove them all in one go.

The combination of pivot table and sparklines gives you great flexibility in exploring your data. In the following exercise you can have a go at using this combination.

Have a Go: Insert sparklines for pivot table data

You'll need the *Butterflies and Site Management.xlsx* dataset for this exercise. The data show the abundance of four species of butterfly at a site in Southern England. The *Qty* column shows the abundance as an index calculated from counts and divided by the length of the transects. There are five different transects and these are subdivided into 12 sections. Some of the sections were managed in various ways and others not. The *Mng* column shows if there was active management or not. The *Year* variable shows the year the data were collected (1996–2007).

↘   Go to the website for support material.

1. Open the spreadsheet; there is only one worksheet, *Data*. You'll notice that there are zero entries for some records (so you should not get gaps). You want to follow the abundances (*Qty*) across the time periods so start by making a pivot table. Click in the block of data then use the *Insert > PivotTable* button and place the result in a *New Worksheet*.

2. Make the pivot table by dragging the fields from the list: *Qty* to *Values*, *Spp* to *Column Labels*, *Year* to *Row Labels*, *Mng*, *Transect* and *Section* to *Report Filter*. The *Qty* field displays "Sum of Qty", which is not what you need so alter this to display the average (mean): click the *Sum of Qty* field and select *Field Settings* and change the summary statistics to *Average*. You can alter the number format too, using the *Number Format* button.

3. Your pivot table should now show four columns of butterfly abundance over several years. You don't need the *Grand Total* items so go to the *PivotTable Tools > Design* menu and use the *Grand Totals* and *Subtotals* buttons to turn all off.

4. Now go to the *PivotTable Tools > Options* menu and in the *Show* section click the *Field Headers* button. Now you are ready to insert some sparklines.

5. Click the *Insert > Line* button to open the *Create Sparklines* dialog box. Click in the *Data Range* box (delete anything there) then select the data (cells B7:E18); you only want the values. Then click in the *Location Range* box and select the cells to hold the sparklines; use cells B20:E20. Click *OK* to insert the sparklines.

6. Now set the row height to about 90: either right-click in the row 20 margin or drag the boundary.

7. Set the column width for the sparklines to about 10; it is easier to highlight all the columns you want and right-click.

8. Now click once on one of the sparklines to activate the *Sparklines Tools*. Click the *Design* tab and then turn on the high and low points in the *Show* section. These do not show up so make the points red; click the *Marker Color* button in the *Design* section and set the appropriate colours.

9. Now click the *Sparkline Color* button (you'll need to click the triangle to open the options) and then select the *Weight* option. Make the lines a little fatter; 1½ points is fine.

10. Click in cell B20, which contains the sparkline for the "pbf" species. Type a formula to place the species name in the cell: =B6 will do nicely. Copy the formula across the other sparklines. Now you should have the names in the cells.

11. Highlight all four sparklines and then go to the *Home* menu. In the *Alignment* section click the buttons to centre the text horizontally and place the text at the top of the cells. In the *Font* section change the font colour to a medium grey. You can also click the bucket icon (*Fill Color*) and add a background

colour (Tan 2 looks good). Now use the *Borders* button and set *All Borders*. Your sparklines should now resemble Figure 6.20.

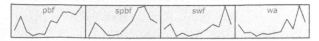

Figure 6.20 Sparklines can be formatted using the *Sparklines > Design* menu. Since they are also regular cells you can set font and background colour and other regular formatting options.

Now use the *Report Filter* items to explore parts of the dataset: the most useful is probably the *Mng* field, look at the different managements by clicking the dropdown triangle icon and selecting items as required.

The *Report Filters* give you a quick way to explore your dataset. The sparklines update each time you apply a filter.

> **Tip: Pivot table autofit and sparklines**
> Each time you alter a pivot table the column widths are updated to autofit the contents. You can turn this off using the *PivotTable Options* (right-click on the table). Untick the option that says "Autofit column widths on update". This means that if you've set wide columns to view sparklines they'll be preserved.

If you want to share sparklines with others you have two main options; save the cells as a PDF or use the clipboard to transfer to another application. You'll see more about sharing charts (including sparklines) in Section 9.3.

Sparklines are useful but you'll need to use regular charts at some stage, since these give a wider range of display options. For time-related data the line plot is the chart of choice; you'll see how to produce these next.

6.2.2 Line plots

You could think of a line plot as being most like a column chart (or bar chart). The line plot shows a single variable at various categorical intervals, which are joined by lines. It is possible to turn the line off; you saw this earlier when making dot charts for error checking (Section 3.3.1). The line emphasizes the link between one interval and the next. Although a line plot looks a bit like a scatter plot there are fundamental differences: the scatter plot has an *x*-axis that is a continuous variable. The line plot has an *x*-axis split into categories, evenly spread along the axis, which is why it is most similar conceptually to a column chart.

> **Note: Category axis in line plots and sparklines**
> The intervals on the category axis of a line plot (the x-axis) are equal. So, if your intervals are not equal you need to take care in interpretation. The same goes for sparklines.

It is possible to display more than one variable in a line plot; you simply differentiate the data series using different colours or styles. The line plot works best when the time periods you're using are equal throughout the dataset, since the x-axis is split by time-period. If you have rather unequal time intervals then a scatter plot may be better – you can add a line to a scatter plot to help emphasize the change from one observation to the next.

There are two main routes to producing a line plot:

- From raw data – you simply select the data you need and construct your chart.
- From a pivot table – a line plot linked to a pivot table is a pivot chart.

As with sparklines your dataset will determine which approach is best; if you have single-date data then make a line plot from the raw data. If you have repeated dates, and by implication grouping variables, then a pivot chart is the best route.

Regular line plots
You create line plots using the *Line* button in the *Charts* section of the *Insert* menu. You have two choices for making a line plot:

- Select data then insert a chart.
- Insert a blank chart then add the data.

There are pros and cons to both approaches. If you select the data first you can use the selection outlines to alter the charted data quickly. However, the chart needs to be in the same worksheet as the data. If you build a chart you have more flexibility but you need a few extra clicks. Using named ranges helps in the chart-building process.

Making a line plot from selected data
Selecting the data first can be a quick option; you can always edit the data afterwards. This approach plots your data using a simple index as the x-axis. You can easily edit the x-axis to show the real data later.

The general process works as follows:

1. Navigate to the worksheet containing the data you wish to explore. The data need to be sorted in date order so click once in the block of data then click the *Data > Sort* button to open the *Sort* dialog box.
2. Select the column holding the column you will use for the x-axis and choose to sort from *Smallest to Largest*. If you have an observation number index column this will probably be in the correct order in any case.

3. Now click the column heading of the variable that will form the y-axis to select it.
4. Click the *Insert > Line* button and choose one of the options. If you have a lot of rows you might choose a line without markers; if there are relatively few rows then you might choose the option to show the line markers.
5. The line plot is now created; the x-axis is a simple index. You can specify the appropriate variable to plot on the x-axis later; for the time being you'll see your data as a line with the name of the variable as both a plot title and in the legend.
6. Click on the chart and you should see selection borders around the data in the worksheet. You can drag the outline to a new variable to plot it; the title and legend update to reflect the new data.

You can edit the chart using the various *Chart Tools* menus. To alter the x-axis to reflect the actual time data, rather than a simple index, you need to do the following:

1. Click once on the chart to activate the *Chart Tools* menus.
2. Now click the *Design > Select Data* button to open the *Select Data Source* dialog box.
3. On the right you'll see a section labelled *Horizontal (Category) Axis Labels*. It will probably contain simple numeric values in an index (Figure 6.21). Click the *Edit* button then select the cells that contain the variable you want. If you have named ranges this can save you some time with the mouse (or you can use keyboard shortcuts).

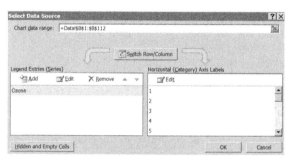

Figure 6.21 The right side of the *Select Data Source* dialog box is where you can edit the x-axis to show your time period variable instead of a simple index.

4. Now you should see the time period labels in the dialog box. Click *OK* to complete the process and update the chart.
5. You can edit the date display by right-clicking on the x-axis and selecting *Format Axis*.

If you edit the x-axis data you'll see that you cannot simply get the selection outlines by clicking on the chart. You need to click on the data line, and then the selection borders become active. You can also go to the *Layout* or *Format* menu and select a data series from the dropdown box in the *Current Selection* section.

Select multiple variables

It is possible to select more than one variable and have them plotted on the same chart. To get a multiple plot you simply click the column heading of the first variable you want then hold the Ctrl key and click the column header of the second variable (the Ctrl key allows you to select non-adjacent columns). Now click the *Insert > Line* button and your line plot will display two lines (Figure 6.22).

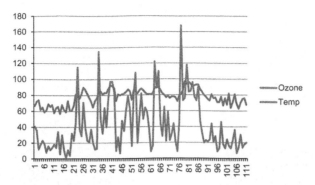

Figure 6.22 Two data series plotted on a line plot. Note that the two variables share the *y*-axis.

If you want to alter the variables plotted you can click one of the lines to bring up the selection outline then drag the outline to a new variable (you'll need to drag the label too). Alternatively you can use the *Design > Select Data* button.

Sometimes one variable is scaled quite differently from the other so you might want to rescale it on a separate axis. You can choose a line and have it displayed on a secondary *y*-axis. Click the line then from the *Layout* menu click the *Format Selection* button. In the *Series Options* section you can choose to *Plot Series On > Secondary Axis* (Figure 6.23).

Figure 6.23 You can choose to plot a data series on a secondary *y*-axis from the *Format Data Series* dialog box.

Once you click *OK* the data are rescaled to fit a second *y*-axis (Figure 6.24).

Once you have the data you want you can use the *Chart Tools* menus to add axis titles and generally edit the chart to make it appear how you want.

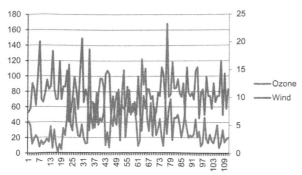

Figure 6.24 Plotting a second variable on a secondary scale can make it easier to see variables with widely differing scales.

> **Tip: Charts and named ranges**
> Having named ranges is particularly useful when you are building and editing charts, as you can select data cell ranges efficiently without having to use the mouse.

Build a line plot from a blank chart

Building a line plot from a blank chart is a flexible approach that gives you good control over the variables you plot. It is especially helpful for large datasets with lots of rows and many variables. Using named ranges is also very helpful in chart-building. In general you proceed as follows:

1. Open the spreadsheet and navigate to the worksheet containing the data. Make sure the data are sorted in time-period order; use the *Data > Sort* button if necessary.
2. Click once in the block of data then use Ctrl+A on the keyboard to select all the data. Now use *Formulas > Create from Selection* to make named ranges, with names based on the top row.
3. Click once in the worksheet to deselect the data. Decide where you want your chart. If you want a new worksheet then make one and navigate to it. If you want to have your chart in the same worksheet as the data you'll need to click on an empty cell that is not adjacent to any data. Using the *View > Freeze Panes > Freeze Top Row* button can be helpful here. You can scroll to the bottom of the data (use Ctrl+Down-Arrow) and insert the chart there.
4. Use the *Insert > Line* button to make a blank line chart; with many rows you'll probably use the plain line option, while with fewer rows you can show the markers.
5. Click once on the chart then use the *Chart Tools > Design > Select Data* button to open the *Select Data Source* dialog box.

6. The section on the left entitled *Legend Entries (Series)* allows you to add variables that will appear on the *y*-axis. The section on the right allows you to select the *x*-axis variable; this defaults to a simple index if you do not select anything.

7. Click the *Add* button to add a new series or *Edit* to edit an existing series. You can then select the cell containing the name (which will be added to the legend), or simply type a name. You can also enter the variable to use for the data; you'll need to add the worksheet name and exclamation mark to use a named range, for example, Data!Ozone.

8. Edit the *Horizontal (Category) Axis Labels* as you need; you can use a named range here.

If you chart a single variable you can use the selection outlines, as long as the chart is in the same worksheet as the data. If you add a second variable, via the *Select Data* button, you can also use the selection outlines but will have to click on a specific line on the chart. You can also alter the data that you view using the *Select Data* button, which works even if your chart is in a separate worksheet.

Once you have a basic line plot you can use the *Chart Tools* menus to help you alter its appearance; many chart elements can also be edited by right-clicking on them.

Pivot chart line plots

A pivot chart is simply a graph that is tied closely to a pivot table. Not all the chart types are available for use as pivot charts; the ones you can use sensibly are:

- Line plots.
- Column charts.
- Bar charts.
- Pie charts.

You can make a pivot chart in two ways:

- Directly: using the *Insert >PivotTable> PivotChart* button.
- Indirectly: make a pivot table first then use the *PivotTable Tools > Options > PivotChart* button.

You've seen how to make pivot tables already to summarize time-related data numerically (Section 6.1.1). For time-related data the line plot is mostly what you are interested in. In general the line plot displays each column as a data series, with each row being a time period and so a point on the line. You can easily rearrange the data in the pivot table of course and any changes will be reflected in the chart. The general way of proceeding is as follows:

1. Navigate to your data and make a pivot table in the usual manner. Place the table in a new worksheet.
2. Arrange your pivot table data so that you have the appropriate columns.
3. Click on the pivot table to activate the *PivotTable Tools* menus.
4. Now click the *Options > PivotChart* button.

5. From the *Insert Chart* dialog box select the chart you want. The dialog box shows all the charts but not all are useable; if you select a chart that cannot be used you'll get an error message (Figure 6.25).

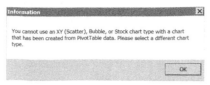

Figure 6.25 Not all chart types are available as pivot charts.

6. Your chart is now created. You can edit and rearrange it as you need.
7. Note that the headings in the *PivotTable Field List* dialog box are subtly different to those used for a pivot table.

You can alter and generally edit your pivot chart in several ways:

- From the *PivotTable Field List* task pane.
- By using filters and the buttons on the chart itself.
- By using the *Slicer* tool.
- By using the *PivotChart Tools* menus, which are broadly similar to the regular *Chart Tools* menus.
- By right-clicking and editing many chart elements directly.

When you make a pivot chart you'll see that there are additional buttons on the chart itself (Figure 6.26).

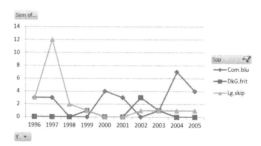

Figure 6.26 A pivot chart includes buttons allowing access to the various fields, for editing and filters.

These buttons give access to the various fields and allow you to alter field settings and apply filters (where appropriate). You can also use the *Slicer* tool to help filter your data; these can be very useful but sometimes it is hard to fit everything on screen (Figure 6.27).
You can access the *Slicer* tool in two ways:

- From the *PivotChart Tools > Analyze* menu.
- From the *PivotTable Tools > Options* menu.

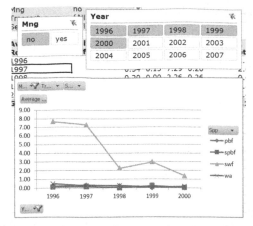

Figure 6.27 The *Slicer* tool can be useful to help filter pivot chart data but screen space can become congested.

In either event your pivot table and chart is updated according to the filters you apply.

If you have missing data your chart may display only some of the time periods, missing out the periods with no data. You can switch on blanks by altering the field settings for the variable that contains the time information; this will usually be the rows of your pivot table. You need to click the field in the *PivotTable Field List* task pane (in the *Row Labels* section) and select *Field Settings*. Then click on the *Layout & Print* tab and tick the box labelled *Show items with no data* (Figure 6.5). Your chart should now show the "complete" time periods but blanks display as gaps (Figure 6.28).

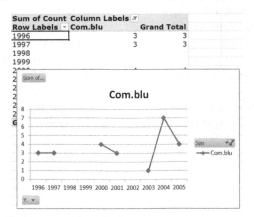

Figure 6.28 Missing data display as gaps by default.

You can alter how the missing data are displayed in two ways:

- Use the *PivotChart Tools > Design > Select Data* button then use the *Hidden and Empty Cells* button from the *Select Data Source* dialog box. You can then choose to display missing data as zero (Figure 6.19).

- Right-click the pivot table and choose *PivotTable Options*. Then choose to display empty cells with a zero (0) by entering a zero in the box labelled *For empty cells show*.

You saw how to deal with missing values earlier when looking at using sparklines (Section 6.2.1).

In the following exercises you can have a go at making some pivot charts and exploring some time-related data. You'll use the butterfly datasets that you met earlier so making the pivot tables should be a familiar process. You'll look at using the Slicer tool to filter and explore data later but in the first exercise you'll need to deal with missing values.

Have a Go: Make a pivot chart for data with missing values

You'll need the *Butterfly and Year Records.xlsx* file for this exercise. The data show the number of various butterfly species (*Spp*) observed at a nature reserve in Scotland. The observations give the *Count*, which is repeated over several years. Note that only positive records are presented: if a species was not observed there is no record for it, in other words there are no zero entries. This is an important point.

↘ Go to the website for support material.

1. Open the spreadsheet; there is only one worksheet, *Data*. There are relatively few time periods (1996 to 2005) but you have a grouping variable, the species (*Spp*, there are 20). Notice that some of the *Count* data are non-integer. The counts were taken from several transects and adjusted for area (lengths of the transects).

2. Click once anywhere in the block of data, then click the *Insert >PivotTable* button to start the process of making a pivot table. The data should be highlighted automatically so choose to place the result in a *New Worksheet* and click *OK*.

3. Drag the *Count* field from the list at the top to the *Values* section at the bottom. Drag the *Year* field to the *Row Labels* section. Now drag the *Spp* field to the *Column Labels* section.

4. Now click once on the pivot table to activate the *PivotTable Tools* menus, then click the *Options > PivotChart* button. The *Insert Chart* dialog box opens and you can select the appropriate chart type. You want the *Line* section; choose one of the options with markers, since there are not many time periods. Click *OK* to insert the chart.

5. Click and drag the chart to a sensible spot so you can see the whole chart and the field list.

6. At the start you'll see all the columns displayed, this is rather too congested so you'll need to use a filter to display fewer items (or even just one). Click the box (labelled *Spp*) above the legend to open the Filter dialog box. Untick the *(Select All)* box and the tick *DkG.frit* to select that single species. Click

OK once you're done and you'll see that there are only data for three of the time periods (years). The pivot table also only shows three rows.

7. Click the *Year* field in the *Row Labels* section of the *PivotTable Field List* task pane and then choose *Field Settings*. Go to the *Layout & Print* tab and tick the box labelled *Show items with no data*. Click *OK* and you'll return to your worksheet. The *x*-axis now shows all the time periods but the missing data show as gaps (similar to Figure 6.28).

8. Now right-click in the pivot table and select *PivotTable Options*. Click in the box labelled *For empty cells show* and type a zero (0), then click *OK* to return to the worksheet. The pivot table will now display zero where before there were blanks, the chart will also now show the zero items (Figure 6.29).

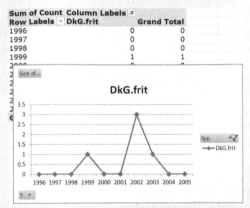

Figure 6.29 You have to alter various settings to display missing data as zeroes in a pivot chart.

9. You can now use the filter to show one or more of the columns (*Spp*); missing data are always displayed as 0.

Practise using the different tools to filter, view and explore the data. If you want a complete reset you can click the chart to activate the *PivotChart Tools* menus then use the *Analyze > Clear* button. The chart and pivot table will be cleared and you can rebuild them.

The *Field Buttons* on the chart are helpful in allowing quick access to the filter tools. They are especially useful if you need to turn off the *PivotTable Field List* task pane, to make more space. The field buttons can be disabled by right-clicking on a button. You can also use the *Analyze > Field Buttons* button. Turning off the field buttons is something you'll probably want to do before sharing your graph with anyone else; you'll see more about sharing graphs in Section 9.3.

The field buttons are useful but they aren't the only way to apply filters. The slicer tool is a good way to manage filters quickly. In the following exercise you can have a go at using a slicer filter.

Have a Go: Make a pivot chart and use a slicer filter

You'll need the *Butterflies and Site Management.xlsx* file for this exercise. The data show the abundance of four species of butterfly on a site in Southern England. The *Qty* column shows the abundance as an index calculated from counts and divided by the length of the transects. There are 5 different transects and these are subdivided into 12 sections. Some of the sections were managed in various ways and others not. The *Mng* column shows whether there was active management or not. The *Year* variable shows the year the data were collected (1996–2007).

↘ Go to the website for support material.

1. Open the spreadsheet; there is only one worksheet, *Data*. You'll notice that there are zero entries for some records (so you should not get gaps). You want to follow the abundances (*Qty*) across the time periods so start by making a pivot table. Click in the block of data then use the *Insert >PivotTable* button and place the result in a *New Worksheet*.

2. Make the pivot table by dragging the fields from the list: *Qty* to *Values*, *Spp* to *Column Labels*, *Year* to *Row Labels*, *Mng*, *Transect* and *Section* to *Report Filter*. The *Qty* field displays *Sum of Qty*, which is not quite what you need so alter this to display the average (mean). Click the *Sum of Qty* field and select *Value Field Settings*. Change the summary statistics to *Average*. You can alter the number format too, using the *Number Format* button.

3. Now click once on the pivot table to activate the *PivotTable Tools* menus, then click the *Options > PivotChart* button. Choose a *Line* style of chart and click *OK* to place it in the worksheet.

4. Click and drag the chart to a convenient spot on screen. You can use the *Field Buttons* on the chart to manage the various filters. You can also access the filters from the pivot table or from the *Field List*. However, for this exercise you'll use the *Slicer* tool. Click the *Analyze* tab then the *Insert Slicer* button to open the *Insert Slicers* dialog box.

5. Tick the boxes by the *Year*, *Spp* and *Mng* items then click *OK* to place the *Slicer* boxes in the worksheet. Click on the top-most slicer to activate the *Slicer Tools* menu (there is only one item, *Options*).

6. Use the mouse to resize and position the boxes in turn. Use the *Columns* button to alter the number of items shown in each box if you need to make a box wider. It is usually best to place the boxes above the chart because its width can alter and move the boxes out of view.

7. Click on the chart then turn off the *PivotTable Field List* task pane using the x at the top of its window. Now, right-click one of the field buttons on the chart and select *Hide All Field Buttons on Chart*. Your worksheet should now resemble Figure 6.30.

8. Use the *Slicer* windows to display the data in various ways. You can select multiple items by holding the Ctrl key whilst clicking. The Shift key selects all items between the item already selected and the item you Shift+Click on.

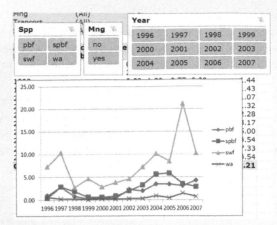

Figure 6.30 *Slicer* windows arranged to allow data exploration. *Field Buttons* and the *Field List* have been turned off.

9. When you have finished experimenting with the slicer filters clear them all using the funnel-and-cross icon in each window. All the data should now be displayed. Click on the chart and use the *Analyze > Field List* button to restore the *PivotTable Field List* task pane.

10. Now drag the *Mng* field from the *Report Filter* to the *Legend Fields (Series)* section, place the field under the *Spp* field that is already there. Turn off the *PivotTable Field List* task pane and look at the chart. You'll see that there are twice as many lines, as each *Spp* item is now split by the *Mng* field (either *yes* or *no*). Use the *Spp* slicer to compare the time series for each species with the different levels of management (Figure 6.31).

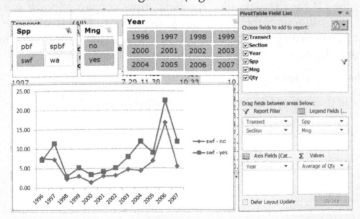

Figure 6.31 Multiple legend series used to compare time-series data.

Slicers are useful for exploring your data; once you've got the slicers set up you can apply the filters quickly and easily.

> **Tip: Slicers and many variables**
> If you have a lot of variables it can be hard to get them all on screen at the same time. Moving the chart to a new window and arranging the screen can help you see the slicers and the chart at the same time, especially if you zoom out the window containing the slicers.

Line plots are the most logical choice when visualizing your time-related data. Pivot tables and pivot charts are especially helpful when you have grouped data, as you've seen. There are many ways you can tweak your charts, you'll see more about formatting charts when you look at sharing your graphs (Section 9.3).

6.3 EXERCISES

1. When you make a sparklines chart any missing values can only be shown as gaps. TRUE or FALSE?
2. On a line plot the intervals on the category axis are ____.
3. Which of the following statements about the *Slicer* tool is true?

   A) You use the *Slicer* instead of the *Report Filter*.
   B) You use the *Slicer* instead of the *Field Buttons*.
   C) You can filter on multiple fields (variables).
   D) You use the *Slicer* instead of the filters in the *PivotTable Field List*.
   E) You can only use a *Slicer* filter instead of the regular *Filter* tools.

4. The *Field Buttons* on pivot charts simply allow you to choose one or more groups for that variable. TRUE or FALSE?
5. Look at the *UKgas.xlsx* data file. The data show consumption of gas in the UK (variable *gas*, millions of therms) for the period 1960 to 1986 (variable *yr*). The data are split into quarters (variable *qtr*). Quickly explore the gas consumption by time of year and the general trend. What are your initial conclusions?

↘ Go to the website for support material.

The answers to these exercises can be found in Appendix 1.

6.4 SUMMARY

Topic	Key points
Time-related data	Data can have a time-sensitive component. The underlying data type can be differences or correlation.
	If you have a differences kind of dataset you can summarize by time period.
	If you have a correlation kind of dataset you may be looking for relationships between variables across the time periods or you may use the time itself as a variable.
Pivot tables	These allow you to arrange your data in time-order. You can use grouping variables to summarize and follow data across time intervals.
	You can place more than one variable in a section of a pivot table, which allows you to explore multiple grouping variables (predictors).
Filters in pivot tables	Report filters allow you to view groups of data in your pivot table.
	You can apply filters to any of the variables in the pivot table, even if the variable is not in the report filter.
	The *Slicer* tool allows a quick way to apply simple filters, especially useful for grouped data.
Sparklines	Sparklines are mini in-cell charts. They are a useful summary for time-related data, especially the line style of sparkline.
	You can use sparklines tools to highlight the high and low points.
	Sparklines are especially useful in conjunction with pivot tables but you can use them on raw data too (for example, following an ungrouped variable across time).
Line plots	A line plot shows the value of a variable (as points) at various categories. For time-related data these categories are the time intervals. The categories are given equal space on the category axis so if your time intervals are unequal you need to take care.
	The points are joined by a line, which allows you to follow the variable across time.
	You can make line plots from raw data or from a pivot table as a pivot chart.
	Multiple variables can be plotted on one line plot, each data series having its own points and line. A data series can be plotted on a secondary *y*-axis, which is useful if there is a large difference in scale.
Pivot charts	You can make line plots as pivot charts, which allow you great flexibility in exploring your data. As well as using report filters and regular filters from the pivot table you can use field buttons right on the chart itself to apply filters.
	The *Slicer* tool can also be used to access filters in a quick and interactive manner.

7

EXPLORING ASSOCIATION DATA

When you are looking for associations your data generally take a particular form: counts of items that fall into particular categories. The end result is usually shown as a *contingency table* (Table 7.1).

Table 7.1 Contingency table showing the frequency of polypore fungi in mangrove swamps.

| | Mangrove type | | |
Fungi	Black	Red	White
Cer.ala	1	0	0
Cer.ane	0	4	0
Dat.cap	0	1	25
Phe.adh	0	0	6
Phe.cal	0	0	1
Phe.gil	0	0	1
Phe.spp	0	1	0
Phe.swi	48	0	0
Tri.bif	1	25	1

The values in the contingency table show the number of items in each of the combinations of categories; this is the frequency. In Table 7.1 you can see the frequency of different species of polypore fungi in mangrove swamps. There are nine categories of fungi (the species) and three categories of mangrove.

You are not likely to have collected your data as a contingency table; it is possible to do but not helpful (so don't!). You will most likely have data arranged in columns, where each column represents one of the categories. You may have a separate column for the frequency or you may not; it is possible to compute the frequencies from the data, as you'll see shortly.

Because association data are not replicated data you cannot use averages and the usual ways of summarizing numerical samples. Your contingency tables become the important features of your explorations and graphical summaries are the most important way to look at your data, as you'll see shortly (Section 7.2). Before that you'll see how to prepare and manage the contingency tables you'll need.

7.1 CONTINGENCY TABLES FROM YOUR DATA

The standard scientific recording format is powerful and flexible; data in this form can be rearranged to make the contingency tables you need easily and efficiently. There are three main forms that your association data can take (Figure 7.1):

- *Contingency table* – this is the final form that you are aiming for; you might receive data in this form or choose to share your data like this.
- *Full scientific records* – this is the most useful format, where each row is an individual observation. You compile the frequencies from the individual records. You may have a column for the frequency (all entries would be 1) but this is redundant, since the category entry is all that is needed. You might like to think of this as long-record format, as opposed to the next form.
- *Compiled frequency records* – this is where the frequencies have been compiled from the original observations. Think of it as short-record format. Each row represents a unique combination of categories and you have a column for the frequency information.

	A	B
1	Species	Mangrove
2	Cer.ala	Black
3	Phe.cal	White
4	Phe.gil	White
5	Phe.spp	Red
6	Dat.cap	Red
7	Tri.bif	White
8	Tri.bif	Black
9	Phe.swi	Black
10	Phe.swi	Black
11	Phe.swi	Black
12	Phe.swi	Black
13	Phe.swi	Black
14	Phe.swi	Black

	A	B	C
1	Species	Mangrove	Freq
2	Phe.swi	Black	48
3	Tri.bif	Black	1
4	Tri.bif	Red	25
5	Tri.bif	White	1
6	Dat.cap	Red	1
7	Dat.cap	White	25
8	Phe.adh	White	6
9	Cer.ane	Red	4
10	Cer.ala	Black	1
11	Phe.cal	White	1
12	Phe.gil	White	1
13	Phe.spp	Red	1
14			

	A	B	C	D
1	Species	Black	Red	White
2	Cer.ala	1	0	0
3	Cer.ane	0	4	0
4	Dat.cap	0	1	25
5	Phe.adh	0	0	6
6	Phe.cal	0	0	1
7	Phe.gil	0	0	1
8	Phe.spp	0	1	0
9	Phe.swi	48	0	0
10	Tri.bif	1	25	1
11				
12				
13				
14				

Figure 7.1 Association data in three forms: long records, short records (with frequencies) and contingency table.

If you already have contingency table data you don't need to do much else in the way of numerical summary! You may wish to add row and column totals, and a grand total to summarize the total number of observations. This is very easy to do using the SUM function.

If you have data in scientific recording format (long or short) you'll need to create contingency tables from scratch; this involves using the *PivotTable* tool, which you've seen extensively already. The general way to proceed is as follows:

1. Navigate to your data, then click once in the block of data.
2. Click the *Insert >PivotTable* button; choose to place the result in a new worksheet and click *OK*.

3. Drag the fields from the *PivotTable Field List* to the appropriate boxes. Usually you'll have two columns representing categories; use one for the *Row Labels* and one for the *Column Labels*.
4. If you have a column for frequency (that is you have short records), drag this to the *Values* section. If you have long records (no frequency variable) you can drag one of the category fields from the list at the top to the *Values* section. You'll end up with one field in two different boxes, but that is fine.
5. If you have any grouping variables drag them into the *Report Filter* section.
6. You probably want to display missing items as zero so right-click the pivot table then select *PivotTable Options*. In the *Format* section tick the box labelled *For empty cells show* then type a zero (0) in the box.
7. There are other formatting options you can choose; the *PivotTable Tools* menus provide various ways to modify and present your pivot tables.

Once you have a basic contingency table you can use filters and drag the fields to different sections to display the table in different, potentially useful, forms. You can use the *Slicer* tool to help apply filters, as you saw when dealing with time-related data (Section 6.2).

There is little difference between the approaches you need when using long or short record data. The main difference is that in short format you have a variable for the frequency, which you can use directly. In the following exercise you can have a go at making a contingency table from long records. Later you'll use short records and see how to view data graphically (Section 7.2).

Have a Go: Make a contingency table from long records
You'll need the *Mangrove Fungi.xlsx* data for this exercise. These data show the incidence of polypore fungi in three different sorts of mangrove forest in the West Indies. There are three worksheets, showing the data in the three main forms: long records, short records and contingency table. The worksheet names are fairly self-explanatory.

↘ Go to the website for support material.

1. Open the spreadsheet and navigate to the worksheet *Long records*; you'll see two columns: the one labelled *Species* contains an abbreviated name of the polypore fungi, and the other contains the names of the mangrove forest types. There are nine species of fungi and three forest types. Now click once anywhere in the block of data, then click the *Insert >PivotTable* button. The data should be selected automatically so choose to place the table in a new worksheet and click *OK*.
2. There are only two fields in the *PivotTable Field List*, one for each column. Drag the *Species* field to the *Row Labels* section and the *Mangrove* field to the *Column Labels* section.
3. You now need to calculate the frequency information but there are no spare fields. Drag the *Mangrove* field from the list at the top to the *Values* area. The

field reads *Count of Mangrove* and the frequencies appear in the table. Note that you could have used the other field; try it and see (but not both at the same time).

4. The table now shows the frequencies and the totals but there are blanks where the frequencies are zero. Right-click in the table and select *PivotTable Options*, then type a 0 in the box to display empty cells. Click *OK* to return to the table; the formerly blank items now show zero.

You now have a basic contingency table. Try moving fields around and using a few filters. The *PivotTable Tools* menu items provide a range of tools for helping you to view your table.

In the preceding exercise you used the long-records format, where there was no frequency data. For an additional exercise you could have a go at the short-record format contained in the same dataset.

Contingency tables are useful because they are the foundation for the statistical analyses you'll need to conduct. They are not so good at summarizing the situation though, especially when you have a large table. This is when a visual summary comes into play. In the following section you'll see how to use pivot charts to build on your contingency tables and provide a mechanism for exploring the data.

7.2 VISUALIZING YOUR CONTINGENCY TABLES

When you have association data your best means of summarizing them is graphically. The starting point is a contingency table and since you're likely to have used a pivot table to make this the sensible option is to use a pivot chart. When summarizing contingency tables there are two main options:

- Column and bar charts.
- Pie charts.

You can make both kinds quite easily from pivot tables. The pie chart has rather fallen out of fashion and is less generally useful for scientific data. The human eye is not good at interpreting angular data and pie charts are therefore to be generally avoided. However, there are occasions when you might want one so you'll see how to make them later (Section 7.2.2).

Column and bar charts are generally more useful than pie charts. If you can display the information as a pie chart then you can certainly use a bar chart instead. Indeed column and bar charts have various options that you cannot get with pie charts, as you'll see next.

7.2.1 Column and bar charts

A bar chart is the general name used to describe a graph where categorical data are displayed by the height of various columns (the bars). Most people think of a bar chart as having vertical bars but you can get charts with horizontal bars too. Excel differentiates between the two options by calling the charts with vertical bars *column charts* and the charts with horizontal bars *bar charts*. Here you'll see the terms used interchangeably. There is no difference in the concept; it is simply the orientation of the bars that changes.

There are two main sorts of bar chart:

- Multiple sample charts.
- Stacked sample charts.

In the first case each bar represents a single sample, so you have as many bars as there are samples (hence the name multiple sample chart). In the stacked sample chart each bar represents a sample, which is then split into subsamples. In other words you have a grouping variable. In general stacked bar charts are not so easy to interpret as multiple sample charts but there are occasions when they can be useful. You'll see stacked charts shortly but first you'll look at how to explore your data using multiple sample charts.

Multiple sample charts
The basic chart for exploring association data and contingency tables is the multiple sample bar chart. Each bar is a separate sample and shows the frequency of observations in that sample (Figure 7.2).

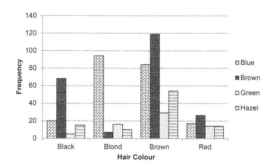

Figure 7.2 Multiple sample bar chart. Each bar shows the frequency of observation from a single sample. This chart shows frequency of hair and eye colour among students at the University of Delaware.

You can easily insert a bar chart from a pivot table; simply click on the pivot table and then use the *Options > PivotChart* button. Once you have a basic chart you can use the *PivotTable Field List* task pane to rearrange the variables; the chart updates immediately. You can apply filters, including the *Slicer* tool, to help you explore your dataset. It doesn't matter whether you use the column or bar chart type of graph; the only difference is the orientation of the bars.

Stacked sample charts

A stacked sample chart is one in which some of the bars have been stacked on one another in a logical grouping (Figure 7.3).

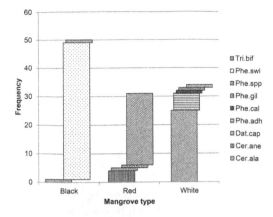

Figure 7.3 Stacked sample bar charts reduce the number of bars but patterns can be hard to differentiate. Offsetting subsamples can help to view the patterns in the data.

This reduces the number of bars but the bars are split into parts, each being a subsample. This is sometimes useful, if you want to compare the composition of groups for example. However, it can be hard to see patterns so offsetting can be used to stagger the subsamples a little (as in Figure 7.3).

There are two sorts of stacked bar chart:

- Regular stacked chart – here the subsamples are simply piled onto one another.
- Percentage stacked chart – the subsamples are converted to percentages, so the frequency axis becomes fixed in the range 0–100%, meaning all bars are the same height.

You can create stacked charts when you make your pivot charts; the *Bar* and *Column* types are split into subtypes, so you can choose the stacked or 100% stacked versions easily. The stacks are formed from the *Axis Fields* section of the *PivotChart Tools* menu, with the subsections being formed from the *Legend Fields* section. Of course it is easy to rearrange the items in the dialog box to shuffle the data and create a different looking chart. You can use filters, including the *Slicer* tool, to help you view portions of your dataset.

Using column and bar charts

In general you want your charts to help you visualize your data and so you should try to minimize clutter. The more variables and groupings you've got the more important it is to use filters to show portions of the dataset, rather than trying to use one chart for everything.

In the following exercises you can have a go at exploring some association data for yourself. The first exercise uses the mangrove fungi dataset, which contains zero frequencies.

Have a Go: Explore association data using bar charts

You'll need the *Mangrove Fungi.xlsx* data for this exercise. These data show the incidence of polypore fungi in three different sorts of mangrove forest in the West Indies. There are three worksheets, showing the data in the three main forms, *Long records*, *Short records* and *Contingency table*. The worksheet names are fairly self-explanatory.

↘ Go to the website for support material.

1. Open the spreadsheet and navigate to the *Long records* worksheet. There are two columns, one for the name of the fungi species and another for the type of mangrove forest.

2. Make a basic pivot table as a contingency table; start by clicking once in the block of data, then use the *Insert >PivotTable* button. The data should be selected automatically so choose to send the result to a new worksheet and click *OK*.

3. Now complete the pivot table by dragging the fields from the box at the top to the sections at the bottom as follows: *Species* to *Row Labels*, *Mangrove* to *Column Labels*, *Mangrove* (again) to *Values*. You should now have a basic contingency table. There are blanks where the frequency is zero: you'll deal with that later.

4. Click once in the pivot table to activate the *PivotTable Tools* menus, then click the *Options > PivotChart* button. Choose the basic column chart from the *Insert Chart* dialog box then click *OK* to create the chart.

5. Now use the field buttons to apply some filters. The most useful button is the *Legend* button, where you can select one or more of the mangrove types to display. When you've had a look at the effects of the filters, clear them all.

6. Now return to the legend *Field Button*, click the button then the tick-box by the *Black* item, to turn it off. You should now be displaying data for two of the three mangrove types. Notice that items with zero frequency are not displayed. This makes the chart more readable but also harder to compare with other charts (since the *x*-axis is different). Click on the *Species* field in the *Axis Fields* section of the *PivotTable Field List* task pane. Select *Field Settings* then go to the *Layout & Print* tab. Tick the box labelled *Show items with no data* then click *OK* to return to the worksheet. Now the *x*-axis shows all the species, which is useful if you want to compare charts.

7. Click on the legend *Field Button* and clear the filter to display all the mangrove types. Now click the *Field Button* for the *Species* (it is in the bottom corner) and bring up the filter dialog. Click the *(Select All)* box to clear all items then enable the ticks on the first two items (*Cer.ala* and *Cer.ane*). Click *OK* to return to the chart. You'll see that only the "Black" and "Red" mangrove types are shown.

8. Enable the showing of empty data in a similar manner to step 6. Click the *Mangrove* field in the *Legend Fields* section and choose *Field Settings*. Now tick the appropriate box in the *Layout & Print* tab. Click *OK* once you've done this to return to the chart. There is not a lot of difference, the items

enabled do have zero value after all, but now there is space on the *x*-axis for the items, which permits an easier comparison with other charts. Use the *Species* field button to look at other combinations of species. Clear the filter once you are done.

9. Now drag the *Mangrove* field from the *Legend Fields* section to the *Report Filter* section. Use the filter to view each mangrove type in turn. Don't forget that you are showing species with zero data so the *x*-axis shows labels for all the species in the dataset. You can alter this using the *Field Settings* as you did in steps 6 and 8. Try disabling zero items and see the effects. Re-enable zero items.

10. Drag the *Mangrove* field from the *Report Filter* to the *Axis Fields* section. You now split the axis into groupings. The order that the fields are in in the *Axis Fields* section has an effect; the field at the top gives the grouping with the field at the bottom being repeated for each level of the other field.

11. Now drag the *Species* field from the *Axis Fields* section to the *Legend Fields* section. Your chart now shows each mangrove split by species. Alter the chart to a stacked type to show the composition for each mangrove: click the *Design > Change Chart Type* button then select a *Stacked Column* type to do this.

12. Now alter the chart type to a *100% Stacked Column*. Your chart now shows bars that extend to 100%, so you are looking at the relative frequencies rather than the absolute. Click the *Layout* tab on the ribbon and select one of the data series using the dropdown box in the *Current Selection* section (you can also right-click one of the data bars). Now click the *Format Selection* button. Use the slider (or type a value in the box) to alter the *Series Overlap*; a value of about 85% is right.

13. Clear the current chart; use the *Analyze > Clear* button. Recreate the original chart (step 3). Now click the *Analyze > Insert Slicer* button and choose both *Species* and *Mangrove* as slicer fields. Rearrange the *Slicer* windows at a convenient spot and use the *Slicer Tools* to make extra columns, which help in fitting the windows on screen. Explore the effects of using the slicers; note that when items are not shown the items are greyed out in the slicer window (Figure 7.4).

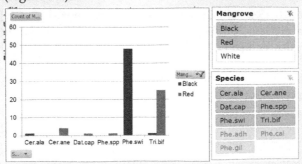

Figure 7.4 Items not shown by the current *Slicer* filter are shown greyed out in the appropriate slicer window.

In the preceding exercise you used the column chart but you could easily use the bar chart instead, the only difference being that the bars are horizontal. Sometimes a horizontal orientation makes more sense for your visual display.

In the next exercise the data includes a grouping variable; this does not make much practical difference to the ways that you can visualize the dataset but it does give you an extra field to utilize.

Tip: Pivot table report filter
The pivot table report filter can be a useful tool to help split and rearrange your data when used with a grouping variable. It is possible to have more than one grouping variable in the filter, so you can get fine control over the results.

Have a Go: Explore association data with a grouping variable
You'll need the *HairEyeColor.xlsx* file for this exercise. The data show the hair and eye colour for students at the University of Delaware in 1974. The data are split into two groups, male and female. There are no zero frequencies in this dataset, so you won't need to alter settings to display empty items.

↘ Go to the website for support material.

1. Open the spreadsheet; there is only one worksheet. There are four columns; the two main variables are *Hair* and *Eye* (for hair and eye colour). The *Sex* column is a grouping variable and the final column contains the frequencies. These data are in short-record format. Click once anywhere in the block of data then use the *Insert >PivotTable* button to start the process of making a pivot table. The data should be selected automatically so choose to place the result in a new worksheet and click *OK*.

2. Make a basic table by dragging the fields in the *PivotTable Field List* as follows: *Hair* to *Row Labels*, *Eye* to *Column Labels*, *Freq* to *Values*. You now have a basic contingency table with male and female results combined in one table.

3. Now drag the *Sex* field item into the *Row Labels* box. Try to get the *Sex* field above the *Hair* field. If they end up the other way simply drag the upper to a position lower down. You now have two contingency tables, one atop the other (Figure 7.5).

4. Try dragging the *Sex* field from the *Row Labels* box into the *Column Labels* box. The result is best if the *Sex* field is above the *Eye* field.

5. Now drag the *Sex* field item out of the *Column Labels* box and into the box entitled *Report Filter*. You now have a single table that combines groups once more. However, you now have a way to manage which groups are displayed (and so combined) using the *Report Filter*.

3	Sum of Freq	Column Labels				
4	Row Labels	Blue	Brown	Green	Hazel	Grand Total
5	Female	114	122	31	46	313
6	Black	9	36	2	5	52
7	Blond	64	4	8	5	81
8	Brown	34	66	14	29	143
9	Red	7	16	7	7	37
10	Male	101	98	33	47	279
11	Black	11	32	3	10	56
12	Blond	30	3	8	5	46
13	Brown	50	53	15	25	143
14	Red	10	10	7	7	34
15	Grand Total	215	220	64	93	592
16						

Fields panel: ☑ Hair ☑ Eye ☑ Sex ☑ Freq

Drag fields between areas below:
▽ Report Filter | ▦ Column Labels: Eye ▾
▦ Row Labels: Sex ▾ Hair ▾ | Σ Values: Sum of Freq ▾

Figure 7.5 A grouping variable can be used to split a contingency table.

6. In row 1 of your spreadsheet there is a label showing the grouping variable used in the report filter (*Sex*), the groups used (*All*) and a small triangle. Click the triangle to display a dialog allowing you to select the groups (Figure 7.6).

	A	B	C	D	E
1	Sex	(All)	▾		
2	Search		🔎		
3	(All)				
4	Female / Male		Brown	Green	Hazel
5			68	5	15
6			7	16	10
7	☐ Select Multiple Items		119	29	54
8	OK	Cancel	26	14	14
9	Grand Total	215	220	64	93

Figure 7.6 The *Report Filter* allows you to combine and display groups.

7. Click on the triangle in cell B1 and then select the *Female* group for the filter. Once you click *OK* the filter is applied and your table displays results for the female students only. The icon of the report filter changes to look like a funnel (Figure 7.7).

	A	B	C	D	E
1	Sex	Female	▾		
2					
3	Sum of Freq	Column Labels			
4	Row Labels	Blue	Brown	Green	Hazel
5	Black	9	36	2	5
6	Blond	64	4	8	5
7	Brown	34	66	14	29
8	Red	7	16	7	7

Figure 7.7 Once a report filter is applied the icon changes to a funnel.

8. Now click the funnel icon to open the filter options. Select (*All*) then click the *OK* button to clear the filter and display all groups (that is, males and females).

9. Click once in the pivot table then click the *Options > PivotChart* button. Choose a basic *Column* chart from the available charts and click *OK* to place the chart in the worksheet.
10. Try using the field buttons on the chart to filter the data in various ways. These allow you to explore the dataset but they are less efficient than the slicer filter tool. Clear the filters before proceeding to the next step.
11. Click once on the chart then click the *Analyze > Insert Slicer* button. Tick the boxes beside the *Hair, Eye* and *Sex* fields then click *OK* to place slicer windows in the worksheet. Move the slicer windows to a convenient spot on screen and use the *Slicer Tools* to display the fields in multiple columns, which usually helps to fit the windows on screen better. Use the slicer filters to see how the display is affected. Clear them all when you are done.
12. Click once on the chart to open the *PivotTable Field List*. Drag the *Sex* field out of the *Report Filter* section and see how it affects the chart when you place it in the *Legend Fields* and *Axis Fields* sections. The order is important, so examine how the chart alters when you place the *Sex* field above or below other fields.
13. Now drag the *Sex* field into the *Axis Fields* section and ensure that it is above the *Hair* field. You now have two blocks of data, effectively splitting the data into the two groups so that you can see the male and female results together in one chart (Figure 7.8).

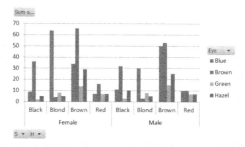

Figure 7.8 When you have a lot of bars your chart can become hard to interpret.

14. Click the chart then go to the *PivotChart Tools > Layout* menu. In the *Current Selection* section (on the left) click the dropdown icon next to *Plot Area* and choose one of the data series. Then click the *Format Selection* button. In the *Series Options* section alter the *Series Overlap* by moving the slider or typing a value in the box; a value of about 50% is about right. Click *OK* to return to the chart. Now the bars overlap one another; this can help you fit more items into your chart but it can also increase the congestion (Figure 7.9).

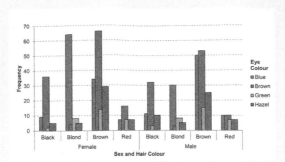

Figure 7.9 Overlapping bars can help fit more categories into a chart but can also increase congestion.

You can use the *PivotChart Tools* menus to help you format your charts; you'll see more about preparing charts when you look at ways of sharing your data (Section 9.3.2).

> **Tip: Labelling chart legends**
> Use the *Layout > Text Box* button to add a title to a chart legend.

The examples you've seen so far assume that you have to make a contingency table using a pivot table. You can of course make a chart directly from your raw data if it happens to be in the form of a contingency table, as you'll see next.

Bar charts from fixed contingency tables
A pivot table gives you maximum flexibility to rearrange your data. However, you may have a fixed contingency table. What you have to do is to insert a regular chart, using the contingency table as the source data. You'll lose some flexibility because you cannot rearrange your table easily but you can still use filters to some degree, as you'll see shortly.

You've got two ways to proceed:

- Make a blank chart, and then select the data you want.
- Let Excel choose your data.

In most cases making a blank chart is the best option because your data are rarely in a form that Excel recognizes properly. However, if you have a contingency table then Excel **can** recognize how the data are arranged. Either way, once you have your chart you can use the selection outlines to move and alter the data range. You can also filter the data using regular filter options. The general way to proceed is as follows:

1. Make sure the contingency table is arranged appropriately. You want any grouping variables on the left, plus the main category variable. Make sure all columns have headings. Subsequent columns relate to the levels of the other category and contain the frequency values. If you have grand totals you should remove them. Your final table might resemble Table 7.2.

Table 7.2 A contingency table prepared for chart insertion. Note that the top row contains headings.

Sex	Hair	Blue	Brown	Green	Hazel
Female	Black	9	36	2	5
Female	Blond	64	4	8	5
Female	Brown	34	66	14	29
Female	Red	7	16	7	7
Male	Black	11	32	3	10
Male	Blond	30	3	8	5
Male	Brown	50	53	15	25
Male	Red	10	10	7	7

2. Now click once anywhere in your contingency table, then click the *Insert > Column* button (or *Bar* if you prefer) and choose a basic multiple sample layout. Click *OK* to place the chart and return to the worksheet.
3. Excel should have recognized the numerical frequency data as the main data series to plot, with the text columns on the left as the category axis (that is, the *x*-axis). If you have any grouping variables they'll be included on the *x*-axis.
4. Click once on the chart to activate the *Chart Tools* menus. You will also see the selection outlines in the data (Figure 7.10).

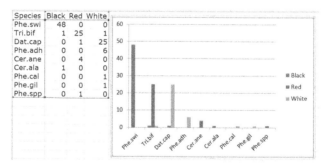

Figure 7.10 Clicking on a chart created automatically displays the data selection outlines, which can be dragged with the mouse to select different data ranges.

5. Use the selection outlines to alter the data that is displayed. You can choose a single column or several but they must be in a single block.
6. Click in the block of data, then click the *Data > Filter* button to enable column filters. You can now use filters to alter the data that is displayed using the dropdown dialog in the column headers. The only sensible filters to apply are the ones for the text columns (the grouping variable and main category).

7. Click on the chart then click the *Design > Switch Row/Column* button. This places the row data as the legend entries and the column data as the category entries (the *x*-axis).

8. You can hide the column containing the frequency data. This also works for row data. Simply right-click on a column (or row) heading (that is, the column letter or row number in the margin) and select *Hide*. Your chart now displays only the rest of the data (Figure 7.11).

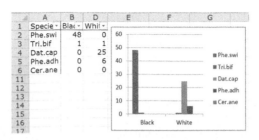

Figure 7.11 If you hide columns or rows their data are removed from the chart. In this case column C and rows 7–10 are hidden. The data cells can be unhidden at any time.

9. To unhide cells you select the visible columns (or rows) around the hidden ones and right-click, then choose *Unhide*.

If you wish to build a chart from a blank worksheet then you'll have to use the *Chart Tools > Design > Select Data* button to choose the data to chart as follows:

1. Click on an empty cell that is not adjacent to any data. Then click the *Insert > Column* button and choose a basic chart.

2. Click once on the chart and then click the *Chart Tools > Design > Select Data* button.

3. You can select the series (the frequency columns) individually or in one block. To select a block click in the *Chart data range* box and highlight the frequency data; include their headings. To choose the data series separately click the *Add* button in the *Legend Entries (Series)* section.

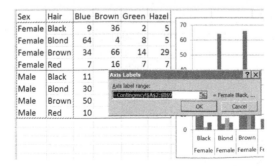

Figure 7.12 Select labels for the *x*-axis with the *Select Data* button. Do not include column headings but you can include a grouping variable if appropriate.

4. In any event you'll need to select the labels for the *x*-axis. In the *Horizontal (Category) Axis Labels* section, click the *Edit* button. Now select the cells containing the labels to appear on the *x*-axis. If you have a grouping variable then select the cells containing those along with the others. Either way, only include the labels and not the column headings (Figure 7.12).

Once you've got a chart you can use all the options you saw previously to filter and explore the data. In summary your filtering options are:

- Use the *Select Data* button and change the charted data.
- Use a regular *Data > Filter* then use the column dropdown dialogs to filter the contingency table.
- Use the selection outlines: click on the chart (or a data bar) and then use the mouse to alter the selection.
- *Hide* and *Unhide* columns or rows.

So, although you cannot use a pivot chart or the slicer tool you can filter your data reasonably well, even when using raw data as a contingency table.

7.2.2   Pie charts

The pie chart remains popular, even though it is not always the best type of visual summary for most data. The pie chart of course is circular, so your data is converted to fractions of the entire circle. Each slice of pie is therefore a percentage and a pie chart shows the relative proportions of the data in question. Because the human eye is not very good with angular data the pie chart is not a great choice. Consider using a regular column chart instead; a 100% stacked column chart does the same job, more efficiently.

You can make pie charts from regular data or as a pivot chart in much the same way as you've seen previously. With most charts the addition of data labels showing the values beside bars of bar charts for example, is unnecessary; the point of the graph is to give a visual summary. With pie charts though, the data labels can be useful. In the following exercise you can have a go at making a pie chart and formatting some data labels.

> **Have a Go: Make a pie chart with formatted data labels**
>
> You'll need the *Mangrove Fungi.xlsx* file for this exercise. These data show the incidence of polypore fungi in three different sorts of mangrove forest in the West Indies. There are three worksheets, showing the data in the three main forms, *Long records*, *Short records* and *Contingency table*. The worksheet names are fairly self-explanatory.
>
> ↘ Go to the website for support material.
>
> 1. Open the spreadsheet. The *Long records* and *Short records* worksheets would need a pivot table to make a contingency table but the *Contingency table* worksheet does not; so navigate to that.

2. Click once in the block of data then click the *Insert > Pie* button. Select a basic pie chart, which is shown immediately.

3. The chart shows the mangrove types bundled together as one sample. Click once in the block of data, and then click the *Data > Filter* button.

4. Now hide the *Black* and *Red* columns (B & C); click in the column header of column B and drag the mouse to select column C too. Right-click and choose the *Hide* option. The columns disappear and the chart updates to show only the fungi species from the *White* mangrove forest type.

5. There are zero frequency items in the data; you can see this from the data. Click the filter dropdown icon in the top row of the *White* column to bring up the filter dialog. You have two choices: the simplest is to untick the box by the "0" item; more complicated is to select *Number Filters*, then choose *Greater Than* and set a value of zero. Either way, apply the filter and return to the chart.

6. At this point the chart is rather rectangular; square shapes work best with pie charts so use the mouse to make the chart outline squarer. Now click the chart and then on the *Design* menu choose a chart style that is all grey colours rather than multi-coloured. You can also use the *Page Layout > Colors* button to alter the colour palette.

7. Now go to the *Chart Tools > Layout* menu and click the *Data Labels* button (Figure 7.13).

Figure 7.13 The *Data Labels* button provides a range of options for placing data labels on charts.

8. There are several options for data labels; choose the *More Data Labels Options* item to open the *Format Data Labels* dialog box (Figure 7.14).

9. Tick the boxes to display the *Category Name*, *Value*, *Percentage* and *Show Leader Lines*. In the *Label Position* section choose the *Best Fit* option. Click *Close* to apply the settings and return to the chart. You'll see some labels in the pie slices and others outside, the latter being ones that don't fit inside the slices.

10. Now you need to do some tidying up: remove the main title (click it then press delete on the keyboard). Now click just outside the pie somewhere on

Figure 7.14 More data labels options from the *Format Data Labels* dialog box.

the lower right, you'll see a selection box showing you've selected the *Plot Area*. Use the mouse to drag this a little lower down the main chart area.

11. Sometimes the in-slice labels do not show up very well; they are usually the ones on dark backgrounds. Click the label *Dat.cap* on the chart; you'll see that all the labels are selected. Click the item again and now only this label is selected. Go to the *Home* menu and make the font white using the *Font Color* button. Repeat the process for the *Phe.adh* label; you should find that this time a single click will select the single label.

12. It can be helpful to highlight slices, especially when you have some large and some small slices. Start by setting the outline for the slices: click in the top corner of the chart then go to the *Layout* menu and use the *Chart Area* dropdown menu in the *Current Selection* section to choose the data series *White*. Then click the *Format Selection* button. Set the *Border Color* to a *Solid Line*. Make the line colour black. You can also set the border to be slightly wider using the *Border Styles* section.

13. Click one of the three small slices near the top. Drag it slightly out from the main pie. Repeat for the other two small slices. Now your chart should resemble Figure 7.15.

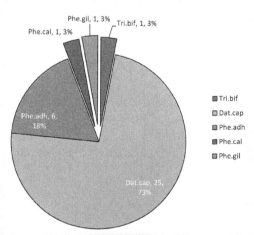

Figure 7.15 Data labels can make a pie chart more readable. Separating some of the smaller slices can highlight portions of your dataset.

In the preceding exercise you based your chart on a completed contingency table. A pivot chart is more easily managed with regard to filtering and arranging of the data than a pie chart; you use the same set of tools for layout and formatting.

> **Note: Data labels on charts**
> You can add data labels to most chart types. Click on the chart then use the *Layout > Data Labels* button. Click any data label to select all labels; click again to select only one label. Thus you can format labels together or singly.

7.3 EXERCISES

1. Association data is about frequencies of observations in ____. You may have data in one of three forms: ____, ____ or a ____.
2. You cannot use any of the traditional measures of centrality with association data. TRUE or FALSE?
3. If you have a fixed contingency table you can only make bar charts. TRUE or FALSE?
4. Which of the following options is **not** available for data labels in a pie chart?

   A) *Percentage.*
   B) *Series Value.*
   C) *Value.*
   D) *Show Leader Lines.*
   E) *Category Name.*

5. Look at the *Titanic.xlsx* data file. The data show the survival (or not) of passengers and crew on the *Titanic*. The data are in short format with frequencies already calculated. Construct appropriate contingency table(s). Do you think that a women and children first policy was successfully implemented?

↘ Go to the website for support material.

The answers to these exercises can be found in Appendix 1.

7.4 SUMMARY

Topic	Key points
Association data	Data are categorical and show the frequency of observations in each category. These are usually summarized in a contingency table, which shows the combinations of items in these categories. The contingency table may or may not also have row and column totals and a grand total.
	Association data are non-replicated; each cell of the contingency table shows a single value (the frequency). So you cannot use averages to summarize the data.

Topic	Key points
Forms of data	Your data can be in one of three forms: Contingency table – frequencies ready assembled. Full scientific records (long records) – each observation is separate and represents a combination of categories. These may be repeated, giving rise to greater frequencies for some combinations. Compiled records (short records) – the frequencies for each category have been combined; each row is a unique combination of categories.
Pivot tables	Use a pivot table to construct a contingency table from long or short records. If you have short records you have frequency data to use in the *Values* section of the pivot table. If you have long records you can place a copy of one of the main categorical variables in the *Values* section to compute the frequencies. If you have grouping variables you can use them to split the table into blocks or use the report filter to view a subset. It is usual to display missing data as zeroes in the contingency table.
Numerical summary	The contingency table is used in statistical analysis and is already a numerical summary. However, a big table can be hard to take in so a visual summary is more helpful.
Bar charts	You can make bar or column charts directly from contingency tables or from a pivot chart based on the long or short records. If you have grouping variables you can display a bar chart with bars beside one another in their groups or as stacked bars. Stacked bars can be displayed to give percentage values (so all bars are the same height). You can offset the elements of stacked bars to make it easier to see the components.
Pie charts	Pie charts show compositional data. Each slice shows the contribution one element makes to the total. A pie chart can only show one category at a time so you may need to present several charts. The human eye is not good at interpretation of angular data so bar charts are generally preferable to pie charts. You can label the slices of pie in various ways.
Filters and slicers	You can use filters on pivot charts in various ways: Report filter – from the pivot table. Regular filters – from fields in the pivot table. Field buttons – directly from the chart. Slicer tool – separate and interactive filter tools. If you make a chart directly from raw data you can still filter using regular filter tools.

8

EXPLORING DIFFERENCES DATA

When you have differences data you need to split your data into chunks, that is, samples based on certain groupings. You might have multiple grouping variables that split the dataset into a hierarchy of samples or you might have a single grouping variable.

In Chapter 4 you saw various ways to summarize samples: numerically, using averages and measures of variability (Section 4.2) and visually, using graphs (Section 4.3). It will come as little surprise to learn that the pivot table is a key element in exploring differences data. The pivot table allows you to group and re-group your data quickly and efficiently. The pivot table also allows you to visualize your data via a pivot chart; you'll see more about pivot charts, and other graphics, shortly (Section 8.2) before that you'll see how to use pivot tables in more detail.

8.1 SUMMARIZING DIFFERENCES DATA

To summarize any sample you need three basic measures:

- *Centrality*, which is a middle value (an average).
- *Dispersion*, which is a measure of variability.
- *Replication*, which is the sample size (how many items you have).

It is also important to know the shape of the data distribution, which determines what measures are most appropriate to summarize the dataset (Section 4.2.4). You saw how to characterize the distribution in Chapter 4 by using shape statistics such as kurtosis and skewness (Section 4.2.4), and how to visualize the distribution using histograms (Section 4.3).

You have two ways to proceed:

- Use filters to display chunks of data in various groupings. Then apply functions to determine the appropriate statistic.
- Use pivot tables to display or summarize the data (you can also use additional functions on a completed table).

You can easily use regular filter operations to display one or more groupings within your dataset. Then you could use any Excel function on the results. The problem here is that your functions point to a fixed range of cells and the filter merely hides the unwanted cells. You can partly overcome this by sorting your data so that data from sample groups lie together in the worksheet. A better solution is to use the *Advanced Filter*, which allows you to send your filter results to a new worksheet (see Section 2.2.3).

You can see readily that using regular filters (advanced or otherwise) is likely to be a tedious business, especially when your dataset contains multiple groups. It is far better and easier to use a pivot table to rearrange your data.

8.1.1   Using pivot tables

A pivot table enables you to arrange and rearrange your dataset quickly and easily. The pivot table contains several built-in functions for summarizing data, for example:

- Sum – simply adds together all the data values.
- Count – how many items there are (the replication).
- Average – the measure of centrality (the mean) for normally distributed data.
- Max – the largest value in the sample.
- Min – the smallest value in the sample.
- Standard deviation – the measure of dispersion for normally distributed data.
- Variance – the standard deviation squared.

There are a few other statistics but these are the most useful. Note that you cannot get the median or quartile values, so if you need non-parametric statistics you will have to arrange your data sample by sample using a pivot table first, and then use the regular functions afterwards. This means that you'll need to use index variables to help you regroup the data in the pivot table.

Using built-in summary statistics
If you can to summarize your data using parametric statistics (that is, the samples are normally distributed) you do not need index variables and can use the summary functions built into the pivot table. You can easily use your response variable more than once, for example, to view the mean and standard deviation: you simply drop the field into place more than once (Figure 8.1).

Figure 8.1 A response variable can be placed in the *Values* area more than once. Each field can be formatted separately.

If you have a response variable in the *Values* area more than once you can format them independently. You start by clicking on a field in the *Values* area and selecting *Value Field Settings*. You can then alter the field from the dialog box that opens (Figure 8.2).

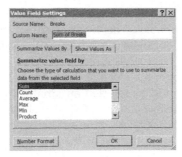

Figure 8.2 The *Value Field Settings* dialog box allows you to change the way a field is displayed.

The most useful things you can alter are:

- The name of the displayed field.
- The type of summary statistic used.
- The format of the result.

Changing the name can be helpful, especially if you are going to share your table with others. You simply type a new name in the *Custom Name* box. You can select one of the summary statistics in the main part of the *Value Field Settings* dialog. Changing the *Number Format* is generally a good idea; you usually don't need more than a couple of decimal places.

Tip: The order of items in pivot tables
Headings in pivot tables are usually set alphabetically. If you right-click on a row or column label you get the option to move the item. This allows you to rearrange rows or columns.

In the following exercise you can have a go at making a table of summary statistics for some data that contains two grouping variables.

Have a Go: Summarize differences data using built-in summary statistics in a pivot table
You'll need the *warpbreaks.xlsx* file for this exercise. The file contains two worksheets: *Recording-layout* contains the basic data and *Recording-layout-indexed* includes extra index columns. The data represent the number of breaks in looms of wool under various tensions. There are two types of wool and three tension settings.

↘ Go to the website for support material.

1. Open the *warpbreaks.xlsx* spreadsheet and go to the *Recording-layout* worksheet. There are three columns: *Breaks* contains the response variable, the number of breaks per loom. The other two columns, *Wool* and *Tension*,

contain grouping variables. You won't need any index variables because the summary statistics can be computed using the built-in functions in the *PivotTable* tool.

2. Click once anywhere in the block of data. Then click the *Insert >PivotTable* button. The data should be selected automatically so choose to place the result in a new worksheet and click *OK*. You should now see the *PivotTable Field List* task pane (if not, click on the empty table on the left of the screen).

3. Drag the *Wool* field from the list at the top to the *Row Labels* area at the bottom. Do the same with the *Tension* field; make sure that *Tension* is below *Wool*.

4. Now drag the *Breaks* field to the *Values* area. You'll see the column appear in the pivot table headed *Sum of Breaks*. Repeat the operation twice more so that you end up with three columns, each containing the same information but with subtly different column headings.

5. Click the top-most of the three fields in the *Values* section of the *PivotTable Field List* task pane. Select *Value Field Settings* from the popup menu. Select *Average* from the list of options in the *Summarize value field by* section. Now click in the *Custom Name* box and type something meaningful, Mean will do. Now click the *Number Format* button and alter the format to a number with two decimal places. Click *OK* to return to the *Value Field Settings* dialog box and *OK* again to return to the worksheet.

6. You now have one column neatly formatted to show mean values. Repeat step 5 for the other two fields in the *Values* section. Use *StdDev* as the summary statistic for one field and *Count* for the last field. The last item doesn't need a special number format because it can only contain integer values.

7. Now your pivot table shows the mean, standard deviation and number of replicates for the various groups in the dataset. You don't need any totals so use the *Pivot Table Tools > Design > Grand Totals* button to turn off all grand totals.

8. At the moment you have subtotals; try rearranging the fields in the *Row Labels* section; you'll see the subtotals alter. You can turn the subtotals off using the *Design > Subtotals* button.

	A	B	C	D	E
1					
2					
3	Wool	Tension	Mean	StdDev	Count
4	A	H	24.56	10.27	9
5	A	L	44.56	18.10	9
6	A	M	24.00	8.66	9
7	B	H	18.78	4.89	9
8	B	L	28.22	9.86	9
9	B	M	28.78	9.43	9

Figure 8.3 You can use a response variable in a pivot table multiple times to obtain a summary.

9. From the *Design* menu use the *Report Layout* button and select *Show in Tabular Form* and also *Repeat All Item Labels*. Now go to the *Options* menu and use the *+/- Buttons* button to neaten the display by removing the "-" symbols in the table. Your final table should resemble Figure 8.3.

Try dragging the grouping variable fields to the *Column Labels* area, singly or together, to alter the general form of the summary table. Having the grouping variables in the *Row Labels* area generally works best.

If you need to use other summary statistics you'll have to use index variables in conjunction with your pivot table, which you'll see next.

Using alternative summary statistics

You've already seen that a pivot table can deal with various summary statistics. If you need something like a median or quartile however, then you need to find a different strategy. The pivot table is still the most effective way to rearrange your data, but you'll need index variables to help you achieve this (Section 2.1.1).

Once you have your samples rearranged how you like you can use regular Excel functions on the data in the pivot table. The downside is that if you alter the pivot table your summary statistics are altered. You can fix your results by pasting the results to a new location as follows:

1. Create your pivot table.
2. Insert the summary statistics you want using regular Excel functions. You can place the results underneath or beside the existing table or even in a new worksheet. It is a good idea to include headers that mirror the pivot table.
3. Select the summary statistics you just created in step 2 and copy them to the clipboard.
4. Navigate to a new worksheet and use *Home > Paste > Paste Special*. Then choose to paste only the values. When you open the *Paste* dialog you may also see options to *Paste Values*; you can choose one of these to shortcut the *Paste Special* operation (Figure 8.4).

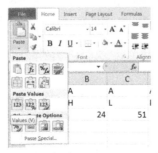

Figure 8.4 Later versions of Excel show multiple paste options, allowing you to shortcut *Paste Special* operations.

5. You can format the result as you need independently of the pivot table. If you alter the pivot table itself the pasted results remain unaltered.

In the following exercise you can have a go at using non-parametric statistics in summary of some samples. You'll have to create a pivot table first, and then carry out the statistical calculations.

Have a Go: Summarize differences data using non-parametric summary statistics via a pivot table

You'll need the *warpbreaks.xlsx* file for this exercise. The file contains two worksheets; *Recording-layout* contains the basic data and *Recording-layout-indexed* includes extra index columns. The data represent the number of breaks in looms of wool under various tensions. There are two types of wool and three tension settings.

↘ Go to the website for support material.

1. Open the spreadsheet and navigate to the *Recording-layout-indexed* worksheet. The column headed *W:T* (column G) is a simple combination of the two predictor variables (*Wool* and *Tension*), and *Sobs* (column H) is an index based on that.

2. Start by making a pivot table, click once in the block of data then use the *Insert >PivotTable* button. Choose to place the result in a new worksheet and click *OK*.

3. Now build the pivot table: drag the *Breaks* field from the list at the top to the *Values* section. Drag the *Sobs* field to the *Row Labels* section. Drag the *Wool* and *Tension* fields to the *Column Labels* section.

4. You now have a basic table that shows columns for each sample but you also have grand totals and subtotal columns. Remove these using the *Design* menu (*Grand Totals* and *Subtotals* buttons).

5. In the *Design* menu use the *Report Layout* button and choose the *Repeat All Item Labels* option, which makes sure that there are no blanks in the column headings of the pivot table.

6. Now click in cell A16 and type a heading, Wool, which will be the heading for your summary table. In cell A17 type Tension. In cell B16 type a formula to copy the heading label: =B4 will do the job. You can now copy the formula across the row to fill up to column G. Then copy down to the next row. You should now have a complete set of labels.

7. Go to cell A18 and type some row headings for the summary statistics you're going to calculate. Start with Max, and then continue down with UQ, Median, LQ and Min. You should have labels in cells A18:A22.

8. Now go to cell B18 and type a formula to work out the maximum for the sample in column B, =QUARTILE(B$6:B$14,4) will do nicely. You could use MAX but this is easier to edit. Note that you need to fix the rows in the formula with $.

9. Copy the formula in cell B18 down the column to row 22. Now they all show the maximum value so edit each one by altering the last parameter to give the appropriate statistic. Cell B19 needs the 4 changing to a 3. Cell B20 needs the 4 changing to a 2. Cell B21 needs the 4 changing to a 1. Cell B22 needs the 4 changing to a 0.

10. Select the cells holding the calculations (B18:B22) then copy to the clipboard and paste across to fill columns C:G. You've now completed the calculations and worked out non-parametric summary statistics for the samples.

11. Click the *Home > Insert* button (you need to click on the triangle dropdown icon) and select the *Insert Sheet* option to make a new worksheet. Return to your pivot table and select the summary statistics you just completed (cells A16:G22). Copy the cells to the clipboard.

12. Now navigate to your new blank worksheet and then click the *Home > Paste* button. In the *Paste Values* section choose the second option, *Values & Number Formatting (A)*. You now have a fixed table of summary values for the samples. Note that you could have used the *Paste Special* option and then selected the appropriate option from the subsequent dialog.

Your table of summary values is now independent of the pivot table. If you alter the table later your new worksheet will be unaffected. It is a good idea to delete the rows you inserted under the pivot table in case you alter the table and partially overwrite them (which could cause confusion).

Once you have your summary data in a new worksheet you can edit as you need. For example, the data can be used as the basis for a chart, which you'll see shortly.

> **Note: Over-paste**
> Use *Paste Special* to overwrite any cells that contain functions with plain values. This can be useful if you have functions that refer to other parts of a spreadsheet that may change (such as a pivot table). Copy the cells to the clipboard and then *Paste Special* them back to the same location, replacing functions with plain values.

Using special variables: calculated fields

You've already seen that the *PivotTable* tool has a variety of built-in summary statistical functions, which aren't always adequate. However, the *PivotTable* tool allows you to specify your own functions, but there is a drawback: you can only use functions that operate on a single cell. So, things like the median, which rely on calculations based on a range of cells, are not available.

In spite of this drawback there are some useful things that you can do with these in-pivot calculations, which are called *calculated fields*. A calculated field is essentially a regular Excel formula inside a pivot table field. You set about inserting a calculated field as follows:

1. Make a pivot table.
2. Go to the *PivotTable Tools > Options* menu. Click the *Calculations* button and then select *Fields, Items & Sets*, which opens a new dialog (Figure 8.5).

Figure 8.5 Inserting a *Calculated Field* into a pivot table starts from the *Options* menu.

3. Select the *Calculated Field* option and the *Insert Calculated Field* dialog box opens (Figure 8.6).

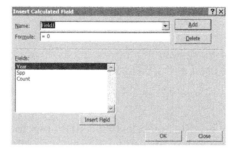

Figure 8.6 The *Insert Calculated Field* dialog allows you to manage calculated fields in pivot tables.

4. Type a name for the new field in the *Name* section.
5. Enter a formula in the *Formula* section. You can use pivot table fields by typing their name or using the *Insert Field* button.
6. Once done click the *OK* button to place the field in the *PivotTable Field List*.

You can edit a calculated field using the same dialog.

Once you have set up a calculated field you can use it like any of the other fields in the *PivotTable Field List* task pane. The new field remains available for as long as you keep the worksheet containing the pivot table.

> **Note: Calculated fields in pivot tables**
> Calculated fields only work on single cells. You cannot create a calculated field that requires a range of cells as an input.

There are many potential uses for calculated fields. In the following exercise you can have a go at making a calculated field that performs a log transformation of the data.

Have a Go: Use a calculated field in a pivot table for logarithmic transformation
You'll need the file *Butterfly and Year Records.xlsx* for this exercise. There is only one worksheet and this contains three columns: *Year* contains the year of observation, *Spp* contains the species name and *Count* contains the recorded abundance.

↘ Go to the website for support material.

1. Start by making a pivot table. Click in the block of data and then use the *Insert >PivotTable* button. Choose to place the result in a new worksheet then click *OK* to go to that worksheet.
2. Put in the headers for the table. Drag the *Year* field from the list at the top to the *Column Labels* section. Drag the *Spp* field to the *Row Labels* section.
3. Now turn off grand totals and subtotals using the buttons in the *Design* menu.
4. Click on the *Options* tab and then *Calculations > Fields, Items & Sets > Calculated Field* (Figure 8.5). The *Insert Calculated Field* dialog box should open (Figure 8.6).
5. Type a name (in the *Name* section) for the new calculated field: Log will do nicely.
6. Now click in the *Formula* section. Delete the 0 to leave the equals sign. Type the formula; you want to end up with = LOG(Count) in the *Formula* section. You can type the name of the field or select it from the *Fields* section and then use the *Insert Field* button. Once you are happy with the formula click *OK* to return to the pivot table.
7. Notice that the field has been added to the list and also entered in the *Values* section. You'll see some errors, #NUM and lots of decimal places. Deal with the decimals first. Click the *Sum of Log* field in the *Values* section then select *Value Field Settings*. Alter the *Number Format* to a basic number (two decimal places is fine). You can also alter the displayed name in the *Custom Name* section, Log(x) will do nicely. Note that you cannot simply use "Log" because a field already exists with that name. Click *OK* to return to the pivot table.
8. Now deal with the error values. Click the *Options* tab then the *PivotTable> Options* button. In the *Format* section tick the box labelled *For error values show:*. Leave the box to the right empty. Click *OK* to return to the pivot table, where the #NUM items are now hidden.

9. You really ought to change the *Log* calculated field because Log(x) is not ideal. Zero values give an error, values of 1 give zero, and values less than 1 give negative results. A better solution is Log(x+1). You'll need to edit your calculated field. Click once in the pivot table then use the *Calculations* > *Fields, Items & Sets* > *Calculated Field* button to open the *Insert Calculated Field* dialog box. The *Name* box will probably display "Field1", so use the triangle icon on the right to see the dropdown list of available fields. There is only one "Log", so select that to bring up its formula. Now click in the *Formula* box and edit the formula to read =LOG(Count +1). Click *Modify* to set the change then *OK* to return to the pivot table.

10. The pivot table now displays a complete table. Items that were previously blank (zero abundance, #NUM) are now showing as 0.00 and items that were previously displayed as 0.00 (abundance = 1) are now showing as 0.30.

Adding 1 to each value is a common trick used when using a logarithmic transformation. The final results are on a log scale, which evens up values that vary across orders of magnitude.

You'll doubtless come up with other uses for calculated fields once you've got the hang of using them.

> **Note: Calculated field formulae**
> Any spaces in calculated field formulae are ignored.

Pivot tables are useful for helping you rearrange your data, even if you then need to carry out further operations. You'll see next how you can use a pivot table to help you prepare data for summary using the *Analysis ToolPak* tools.

8.1.2 Using the *Analysis ToolPak*

You saw earlier how the *Analysis ToolPak* could be used to obtain summary statistics for samples (Section 4.2.5). In most cases you won't have your data arranged like this (in samples) because it is not efficient or flexible. However, the *Analysis ToolPak* can use data in a pivot table as its source so the simplest way to proceed is as follows:

1. Make a pivot table from your data. You'll need to use index variables to rearrange your data into sample groups.
2. Click the *Data* > *Data Analysis* button to open the *Data Analysis* dialog box.
3. Choose the *Descriptive Statistics* option and then click *OK*.
4. Use the mouse to select the data, include any column headings.
5. Tick the options *Grouped By Columns* and *Labels in first row*.
6. Choose to place the output in a *New Worksheet Ply*.

7. Tick the output options you want; *Summary Statistics* is the main option you need.
8. Click *OK* to calculate the results and place them in a new worksheet.

If you have more than one grouping variable you won't be able to incorporate them into your output directly without a little manipulation. The simplest way is to use a new index variable that acts as a label. Look at the *warpbreaks.xlsx* dataset; there are two grouping variables, *Wool* and *Tension*. The *W:T* variable is a simple index created from the first letters of the levels of the variables (see Section 2.1.1). The numerical index variable *Sobs* is calculated from this using the COUNTIF function (Figure 8.7).

	H2					f_x	=COUNTIF(G$2:G2,G2)	
	A	B	C	D	E	F	G	H
1	Breaks	Wool	Tension	Obs	Wobs	Tobs	W:T	Sobs
2	26	A	L	1	1	1	AL	1
3	30	A	L	2	2	2	AL	2
4	54	A	L	3	3	3	AL	3
5	25	A	L	4	4	4	AL	4
6	70	A	L	5	5	5	AL	5
7	52	A	L	6	6	6	AL	6

Figure 8.7 Index variables can help provide labels to use when only a single row of headings is possible.

Once you have the appropriate index variables it is easy to make your pivot table then run the *Analysis ToolPak*. The output is not especially convenient because the headings line up over the columns containing the labels for the summary items, rather than the values. You can deal with this by clicking on the first of the heading labels and using the *Home > Insert > Insert Cells* button. You'll then be able to shift the entire row one space to the right (Figure 8.8).

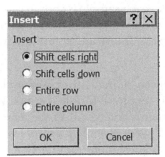

Figure 8.8 Inserting a single cell shifts the existing cells to make room.

Once you've shifted the heading row you can delete the repeated columns containing the labels for the various summary statistics. You can then format the rows as you need, to produce a sensible output (Figure 8.9).

Summary tables are useful but visual methods of summary are easier to interpret. You'll see how to use graphical methods to explore your differences data next.

	A	B	C	D	E	F	G
1	Statistic	AH	AL	AM	BH	BL	BM
2	Mean	24.56	44.56	24.00	18.78	28.22	28.78
3	Standard Error	3.42	6.03	2.89	1.63	3.29	3.14
4	Median	24	51	21	17	29	28
5	Mode	#N/A	26	18	15	29	39
6	Standard Deviation	10.27	18.10	8.66	4.89	9.86	9.43
7	Sample Variance	105.53	327.53	75.00	23.94	97.19	88.94
8	Kurtosis	-0.04	-1.79	-1.57	-0.09	-0.56	-1.46
9	Skewness	0.52	0.16	0.22	0.84	0.30	0.18
10	Range	33	45	24	15	30	26
11	Minimum	10	25	12	13	14	16
12	Maximum	43	70	36	28	44	42
13	Sum	221	401	216	169	254	259
14	Count	9	9	9	9	9	9
15	Confidence Level(95.0%)	7.90	13.91	6.66	3.76	7.58	7.25

Figure 8.9 The output from the *Descriptive Statistics* section of the *Analysis ToolPak* needs some editing to make a coherent table.

8.2 VISUALIZING DIFFERENCES DATA

It is always easier to see patterns in your data when you use a graphical method, compared to using tables of numbers. Later on you'll see how to share your data (Chapter 9), and graphical methods become very important then because they convey a lot of information in a small space.

When you have differences data there are two main types of graph that are useful:

- Bar charts – each bar (horizontal or vertical, what Excel calls a column chart) represents the value for a sample from your dataset (usually the mean). You can add error bars to give a measure of variability.
- Box-and-whisker plots – each sample is shown as a box (which is why the graphs are sometimes called simply *boxplots*), with the limits being the quartiles. The whiskers extend to the maximum and minimum, whilst the median is shown as a point.

Bar charts (or column charts) can be made from a pivot table as pivot charts, but box-and-whisker plots cannot, so require additional steps. When you explore your data for the first time you are certainly interested in the average values for your samples. At some point however, you'll want to look at sample variability. Recall the minimum items you need when presenting a sample summary:

- Centrality – average of some sort (most commonly the mean or median).
- Dispersion – a measure of the variability.
- Replication – how many observations there are in the sample.

With the bar chart you can add error bars to show the variability; the standard deviation is one such measure but standard error is more useful as it incorporates the sample size. You'll see how to make bar charts in Section 8.2.3. With box-and-whisker plots the variability is already incorporated into the plot. You can make box-and-whisker plots using the *Other Charts* button on the *Insert* menu, as you'll see a little later (Section 8.2.4).

You may also want to explore the data distribution of the samples. This will help you to determine which summary statistics are most appropriate and ultimately which statistical test is most appropriate. You saw histograms earlier (Section 4.2.4), when you looked at general ways to make sense of your data. It is easy to make a histogram using a pivot table and the *Analysis ToolPak*, as you'll see shortly (Section 8.2.2).

Sparklines, which you've seen before in conjunction with regression data (Section 5.2.1) and time-related data (Section 6.2.1), can also be used; you'll see these next.

8.2.1 Sparklines

Sparklines are mini-charts that fit inside a single worksheet cell. They can be handy for getting a quick overview of time-related or regression data when used as line charts. However, you can press them into service with a bar-chart version. Sparklines are not an ideal method to view your data (you cannot label the bars) but they are very quick to insert and therefore they are a useful addition to your armoury.

When you have differences data you'll want to view the data by sample groups. This entails using a pivot table to set out your data using the mean values. Means are not always the best average but the median is not available in a pivot table. You could arrange your pivot table to show your sample groups, then compute the medians, and then create the sparklines.

To create your sparklines you can use the following procedure:

1. Construct a pivot table; you want to end up with mean values (averages) so you'll have to alter the default Sum. You can also alter the order of the samples (you can also do this afterwards).
2. Highlight the cells where you want the sparklines to appear. You need to select the correct number of rows or columns, according to the data.
3. Click on the *Insert* tab and then in the *Sparklines* section click the *Column* button.
4. Use the mouse to highlight the data to chart; include only data and not any headings.
5. You already selected the *Location Range* so you should see the cells already named in the dialog. If you did not select the target cells before, you can do that now. Click *OK* to place the sparklines in the worksheet.
6. Highlight the cells with the sparklines to activate the *Sparklines Tools* menu (there is only one section, *Design*). You can use the tools to help format the sparklines.
7. You can also format the worksheet cells, making them wider and taller is often useful. Don't forget that they are real cells so you can add regular text (or a function) to cells containing a sparkline chart.

You can even place sparklines in a pivot table (Figure 8.10) but if you intend to alter the table you should clear the sparklines first.

Sparklines are not ideal for summaries but they can help you identify samples that stand out (by being large or small) from the rest.

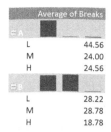

	Average of Breaks
☐ A	■
L	44.56
M	24.00
H	24.56
☐ B	■ ■ ■
L	28.22
M	28.78
H	18.78

Figure 8.10 Sparklines can be placed inside a pivot table. Here the sparkline charts show as bars in the header rows.

8.2.2 Histograms

Histograms are useful because they show you the shape of your data. You need to know this in order to determine which summary statistics are most appropriate. Ultimately you'll also need to decide which statistical test is most appropriate.

You met histograms earlier, when you looked at different kinds of graph (Section 4.3.1). When you have differences data you need to use a pivot table to rearrange your data into sample groups. You can then use the *Analysis ToolPak* to create a histogram. The *Analysis ToolPak* works out the frequency ranges for you automatically, which is useful.

If you need to make a more sophisticated histogram later on you can work out the bin ranges separately for each sample. Alternatively you can use the entire dataset to make one set of bins used by all the histograms.

The general way to proceed is as follows:

1. Make a pivot table. You want to end up with columns representing the various sample groups. This will probably mean you'll need index variables in your dataset.
2. Click the *Data > Data Analysis* button to open the *Data Analysis* dialog box (you'll need to have the *Analysis ToolPak* installed as described in Section 4.2.5).
3. Select the *Histogram* option and click *OK* to open the *Histogram* dialog box.
4. Highlight the cells containing the sample you want to make the histogram for. You do not need to include a header label. If you do include the label make sure the *Labels* box is ticked.
5. If you have a bin range calculated already, specify it in the *Bin Range* box. If not then leave it blank and the *Analysis ToolPak* will work out a basic bin range for you.
6. Choose where to place the result; *New Worksheet Ply* is most useful. You can type a name for the new worksheet if you like (it is a good idea to do so).
7. Now tick the box labelled *Chart Output* (you'll just get a frequency table if this is unticked). Click *OK* to make the histogram (and table of frequencies) in a new worksheet.
8. Edit the chart as you like (see Section 4.3.1).

If you want a more tailored histogram you will need to determine the bin ranges and frequencies for each sample (see Sections 4.2.4 and 4.3.1). This is somewhat time-consuming but there is really no alternative.

8.2.3   Bar charts

Bar charts are generally useful in showing you the values for various sample group-ings. You can easily produce bar charts using a pivot table and pivot chart. You saw this approach when you looked at exploring association data (Chapter 7). For a first look this approach can work quite well as long as you only require the mean as an average. However, if you want to look at the median values you'll need a slightly dif-ferent approach.

It is also important to get an idea of the variability of your samples. For normally distributed data appropriate measures are standard deviation and standard error. For non-parametric data the maximum and minimum values and the inter-quartiles are commonly used. You can add error bars to bar charts (and many other sorts of chart) quite easily but you'll need an extra step or two to achieve the desired result. You reach the tools for making error bars by clicking on the *Chart Tools > Layout > Error Bars* button.

Essentially you've got two ways to make your bar charts:

- Make a pivot table then a pivot chart.
- Make a pivot table then build a bar chart from scratch.

In either case a pivot table is the starting point, as it allows you to arrange your data in sample groupings with minimal effort.

Pivot chart bar charts

Using a pivot chart is the quickest way to get a bar chart from your data. The pivot chart gives you greater flexibility but you cannot easily add error bars; you'll have to do all the calculations separately from the pivot table, so you might as well build a chart from scratch and not bother with a pivot chart. So, use the pivot chart approach for your first look over the dataset so that you can spot the general patterns. Once you've got your pivot table you can modify it easily at a later stage to form the basis for a more sophis-ticated bar chart, as you'll see shortly. To make a basic pivot chart follow this process:

1. Start by making a pivot table. You do not need any index values because you can use the means for the sample groups as your summary statistic. Fields placed in the *Row Labels* area are used on the category axis (the *x*-axis in a regular column chart). Fields placed in the *Column Labels* area form the groupings, which are listed in the legend.
2. Make sure your response variable (in the *Values* area) is set to display the *Average*: click the field then choose *Value Field Settings,* then change the summary statistic to *Average*. This is also the time to alter the *Number Format*.
3. Convert the pivot table to a pivot chart: click the pivot table to activate the *PivotTable Tools* menus, then click the *Options > PivotChart* button.
4. Choose the type of chart you want from the *Insert Chart* dialog box.
5. You can now rearrange the fields in the *PivotTable Field List* to alter the arrangement of the chart.
6. Use the *Field Buttons* on the chart to filter the various variables.
7. Use the *PivotChart Tools* menus to help you alter and format the chart's appearance.

You've already had practice at using bar charts when you looked at exploring associa-tion data (Section 7.2.1). In the following exercise you can have a go at making a bar chart for a differences dataset to help consolidate the skills you gained earlier.

Have a Go: Use a bar chart as a pivot chart to explore differences data

You'll need the *warpbreaks.xlsx* data file for this exercise. The file contains two worksheets; *Recording-layout* contains the basic data and *Recording-layout-indexed* includes extra index columns. The data represent the number of breaks in looms of wool under various tensions. There are two types of wool and three tension settings.

↘ Go to the website for support material.

1. Open the spreadsheet and navigate to the *Recording-layout* worksheet. You do not need any index variables for this exercise. Click anywhere in the block of data then click on the *Insert >PivotTable* button. Choose to place the result in a new worksheet and click *OK*.
2. Construct a basic table by dragging the fields in the *PivotTable Field List* task pane from the list at the top to the sections at the bottom: drag the *Breaks* field to the *Values* section. Drag the *Wool* field to the *Row Labels* section. Drag *Tension* to the *Column Labels* section.
3. The *Breaks* field is the response variable and is showing *Sum of Breaks*. Alter the summary to *Average*: click the field in the *Values* area and select *Value Field Settings*. Set the summary to *Average* and also click *Number Format* to set to a number with two decimal places. In the *Custom Name* box alter the name to read Mean Breaks.
4. Remove grand totals and any potential subtotals: click the pivot table once then use the *Grand Total* and *Subtotals* buttons on the *Design* menu.
5. The *Tension* fields are displayed in alphabetical order but it would be better to have *L, M, H* to represent low, medium and high tension. Right-click in cell B4, which is the cell with the *H* column label. Hover over the *Move* option and a popup menu with more options appears (Figure 8.11).

Figure 8.11 Right-clicking a row or column heading in a pivot table gives you access to the *Move* options, allowing you to alter the order of items in the table.

6. Choose the *Move "H" to End* option to shift this column right to the end. Now the columns are arranged in a more sensible order.
7. Click the *Options > Field Headers* button to neaten the table: it turns off the labels that say *Row Labels* and *Column Labels*.
8. Now click the *Options > PivotChart* button to open the *Insert Chart* dialog box. Choose a basic column chart and then click *OK* to place the chart. You should now see something resembling Figure 8.12.

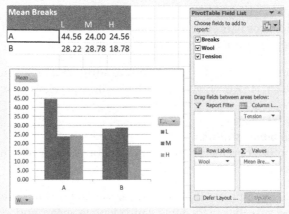

Figure 8.12 A bar chart as a pivot chart. You can rearrange the chart from the pivot table easily. The field buttons on the chart allow access to filters.

9. You can use the *Field Buttons* on the chart to help you filter the data. You can also access the filters from the *PivotTable Field List* task pane (click a field in the top part of the pane). The *Pivot Chart Tools* menus provide tools for modifying the chart's appearance.

Now you have a basic bar chart. Try moving the fields between the four main sections of the *PivotTable Field List* task pane. You can obtain a good overview plot by having both *Wool* and *Tension* grouping variables (predictor variables) in the *Axis Fields* section.

Once you have your pivot chart you can rearrange and filter it very easily, so it makes a good choice for preliminary exploration of your dataset.

Tip: Log scale axis for bars with very different heights
In some datasets the bars on a chart may be very different in size. If your data differs by orders of magnitude it can be hard to see the smaller bars, since the largest bar sets the scale of the value axis. Set the axis to a log scale by formatting the axis. Go to the *Layout* menu then select the appropriate axis. Then use *Format Selection* to open the options. You'll see an option for *Logarithmic scale*; the base defaults to 10.

A pivot chart is useful but at some point you should add error bars so that you can visualize the variability in your samples. You may also need to use a median rather than the mean. In such cases you should begin with a pivot table but rather than make a pivot chart you should build your own chart from scratch, as you'll see next.

Building bar charts

Building your own bar chart gives you more control over the output but the trade-off is that you lose some flexibility, compared to the pivot chart approach. The main advantages are:

- You can add error bars easily.
- You can use non-parametric summary statistics.

The main disadvantage is that you have to create additional table items before you can make a sensible chart. However, you can still use filters to view your data, as you'll see shortly.

Using parametric summary statistics

If you are happy to use parametric summary statistics (that is, the mean and standard deviation) the basic process is as follows:

1. Start by making a pivot table of your sample data. You do not need any index variables because the pivot table can calculate mean and standard deviation directly from the data.
2. Arrange your pivot table so that the grouping variables are in the *Row Labels* section of the *PivotTable Field List* task pane.
3. Drag the main response variable to the *Values* section of the *Field List* task pane.
4. Now click the field in the *Values* section and select *Value Field Settings*. Alter the summary statistic to *Average* and use the *Number Format* button to set the result to a sensible level of precision (that is, not too many decimal places). You can also alter the *Custom Name* displayed in the pivot table.
5. Repeat steps 3 and 4 but set the summary statistic to *StdDev*, which is the standard deviation.
6. You can use standard deviation for error bars but the standard error is a more useful measure, which takes into account sample size. There is no function to compute standard error but you can work it out from standard deviation and the sample size. So, repeat steps 3 and 4 but set the summary statistic to *Count*, which gives the number of observations in each sample group.
7. Now you have a pivot table with a column for each grouping variable (predictor variable) and a column each for mean, standard deviation and sample size.
8. Tidy up the table before you carry out any further calculations. Click once in the pivot table to activate the *PivotTable Tools* menus. Now turn off the grand totals and subtotals using the *Grand Totals* and *Subtotals* buttons on the *Design* menu.
9. Use the *Report Layout* button to set to: *Show in Tabular Form* and *Repeat All Item Labels*. You should end up with a pivot table with appropriate column headings and without any blank labels (Figure 8.13).

Wool	Tension	Mean	StdDev	n
⊟A	H	24.56	10.27	9
A	L	44.56	18.10	9
A	M	24.00	8.66	9
⊟B	H	18.78	4.89	9
B	L	28.22	9.86	9
B	M	28.78	9.43	9

Figure 8.13 A pivot table set out with parametric summary statistics. This can form the basis for a bar chart.

10. Add a column label to the right of the pivot table; you'll place the standard error values underneath.
11. Calculate standard error using: standard deviation / square root (number of observations). So, if your standard deviation is in column D and the count (number of replicates) is in column E your formula would be something like: =D4/SQRT(E4).
12. Copy the formula down the rest of the column and format the cells to a sensible number of decimal places (Figure 8.14).

			F4			f_x =D4/SQRT(E4)	
	A	B	C	D	E	F	G
1							
2							
3	Wool	Tension	Mean	StdDev	n	SE	
4	⊟A	H	24.56	10.27	9	3.42	
5	A	L	44.56	18.10	9	6.03	
6	A	M	24.00	8.66	9	2.89	
7	⊟B	H	18.78	4.89	9	1.63	
8	B	L	28.22	9.86	9	3.29	
9	B	M	28.78	9.43	9	3.14	

Figure 8.14 Standard error calculations added adjacent to a pivot table. The data can now be used to make a bar chart.

13. You can now copy the finished table to a new worksheet. Make a new worksheet then highlight the pivot chart plus the column of standard error values (click once in the block of data then use Ctrl+A), and then copy to the clipboard.
14. Navigate to your new worksheet and click in cell A1. Now use the *Paste* button to display the values and number formatting options (see Figure 8.4); these are also available via *Paste Special*).
15. Now you have fixed values in a new worksheet that you can use as the basis for a bar chart. First click once in the block of data then click the *Data > Filter* button. This allows you to filter the data shown in the chart.
16. Make a bar chart (column chart): click anywhere away from the block of data then use the *Insert > Column* button, and then select a regular column chart and OK to make a blank chart.
17. Use the *Design > Select Data* button to choose the data you want to chart. Click the *Add* button in the *Legend Entries* section. Then use the mouse to select a cell for the name (the column heading for the mean values is adequate), and then the range of cells holding the mean values. Click OK to return to the *Select Data Source* dialog box.

18. Now you need to alter the horizontal axis labels, which currently show a basic index of numbers. Click the *Edit* button then select the range of cells containing the predictor variable labels. If you have more than one column make sure to select them all. You do not need to include the headings, just the labels. Click *OK* a couple of times to return to the worksheet.

19. To make error bars you need to use the *Layout > Error Bars* button. Click this to bring up a host of options (Figure 8.15).

Figure 8.15 Error bar tools are found on the *Layout* menu.

20. Note that the *Error Bars* button appears to give you an option for using standard error. Do not use it; the values it computes are incorrect! Choose the *More Error Bars Options* item to open a new dialog box (Figure 8.16).

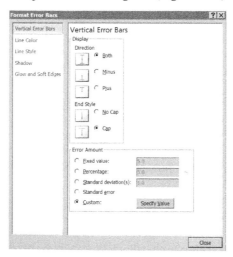

Figure 8.16 The *Format Error Bars* dialog box allows you to select error bar data and formatting options.

21. Click the *Error Amount > Custom* option then click the *Specify Value* button. This opens a new window allowing you to select the error value data. Use the mouse to highlight the cells with the values for standard error. You'll have to add them twice, once for the up bars and once for the down bars. Click *OK* to set the values then use the *Close* button.

Your chart now contains error bars using standard error (Figure 8.17).

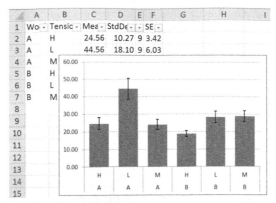

Figure 8.17 A bar chart with error bars (standard error). The data can be filtered using the filter tools.

Once you have a bar chart made in the way just described you can use the filters to display portions of the dataset. This hides some of the rows of data, which might affect the chart size. You can get over this by placing the chart in a separate worksheet or below the table of summary data; whichever is most convenient.

Although you lose a bit of flexibility you can return to your original pivot table and alter that independently of the chart.

Using non-parametric summary statistics

If your data require you to use non-parametric summary statistics (that is median, inter-quartiles) you need a slightly different approach. You'll still require a pivot table but this time you will have to use index variables to arrange your data in samples, column by column (or row by row). You can then use regular Excel formulae to work out the appropriate statistics, as you saw earlier (Section 8.1.1).

When you use non-parametric statistics you need to calculate the median as a measure of centrality. The error bars can be the max–min or the inter-quartile range. In either case you'll need an additional step. The error bars are a deviation from the middle. Put another way, you specify the length of the error bars. This means you need to work out how far above the median value the top and bottom of the error bars will stick out.

Once you have calculated the appropriate values you can build your bar chart. The data can be filtered, allowing you to alter the chart as you saw in the preceding section.

In the following exercise you can have a go at making a non-parametric bar chart. You'll use the inter-quartile range for the error bars.

Have a Go: Build a non-parametric bar chart

You'll need the *warpbreaks.xlsx* data file for this exercise. The file contains two worksheets; *Recording-layout* contains the basic data and *Recording-layout-indexed* includes extra index columns. The data represent the number of breaks in looms of wool under various tensions. There are two types of wool and three tension settings.

↘ Go to the website for support material.

1. Open the spreadsheet and navigate to the *Recording-layout-indexed* worksheet, as you need to use the index variables to help re-organize the data into sample groups.
2. Start by making a pivot table in a new worksheet. Arrange the fields as follows: *Breaks* in the *Values* area. *Wool* and *Tension* in the *Row Labels* area (place *Wool* above *Tension*) and *Sobs* in the *Column Labels* area.
3. Rearrange the order of the *Tension* rows: right-click in the cell containing the "H" label (probably cell A6) then choose *Move* and select to move the "H" item to the end.
4. Click in the pivot table then use the *Design* menu to turn off grand totals and subtotals. Then set the *Report Layout* to *Tabular Form* and *Repeat All Item Labels*. Your pivot table should resemble Figure 8.18.

Sum of Breaks		Sobs									
Wool	Tension	1	2	3	4	5	6	7	8	9	
A	L	26	30	54	25	70	52	51	26	67	
A	M	18	21	29	17	12	18	35	30	36	
A	H	36	21	24	18	10	43	28	15	26	
B	L	27	14	29	19	29	31	41	20	44	
B	M	42	26	19	16	39	28	21	39	29	
B	H	20	21	24	17	13	15	15	16	28	

Figure 8.18 Pivot table set out by sample, ready for calculation of non-parametric summary statistics.

5. Now you can calculate the summary values you require but first make some heading labels. Click in cell L4 and type a formula to copy the label in cell A4 (the column heading for the *Wool* variable), =A4 will do nicely. Copy the formula to the other heading cells so you end up with labels in cells L4:M10.
6. In columns N, O and P you'll calculate the median, upper quartile and lower quartile. In cell N4 type a label Median, then in the adjacent columns type headings UQ and LQ for the other values.
7. The median is the second quartile so in cell N3 type the value 2. In cells O2 and P2 type 3 and 1 respectively (for the third and first quartiles).
8. Now click in cell N5 and type a formula to calculate the median, =QUARTILE($C5:$K5,N$3). Note that you fix the columns for the sample range so that when you copy the formula the correct values are selected. The N$3 part points to the number 2, which means the median is calculated. You fix the row this time, so that when the formula is copied you keep the correct reference.

9. Copy the formula in cell N5 across to fill columns O and P. Now copy down to fill the rows 5:10. You should now have all the appropriate values for the six samples.

10. Click in cell Q4 and type a label for the upper error bar; Up will do nicely. In cell R4 type Dn for the lower error bar.

11. Click in cell Q5 and type a formula to work out the magnitude of the error bar (that is the deviation from the median), =ABS($N5-O5). Add the $ to fix the formula to the column containing the median. Now copy this cell down the rest of the column, then to the adjacent column. You should now have a set of values for both up and down error bars (Figure 8.19).

	*f*ₓ	=ABS($N5-P5)									

G	H	I	J	K	L	M	N	O	P	Q	R
								2	3	1	
4	5	6	7	8	9	Wool	Tension	Median	UQ	LQ	Up	Dn
5	70	52	51	26	67	A	L		51	54	26	3	25
7	12	18	35	30	36	A	M		21	30	18	9	3
8	10	43	28	15	26	A	H		24	28	18	4	6
9	29	31	41	20	44	B	L		29	31	20	2	9
6	39	28	21	39	29	B	M		28	39	21	11	7
7	13	15	15	16	28	B	H		17	21	15	4	2

Figure 8.19 Non-parametric summary statistics calculated from a pivot table, ready to build a bar chart with error bars.

12. You can now copy the data to a new worksheet, using *Paste Special* to preserve the values and formatting, or you can keep them in place. In any event select the summary data, including the headings, then click the *Data > Filter* button. You'll now be able to filter the data displayed in the chart.

13. Now click anywhere not in or adjacent to any data. Then click *Insert > Column* and create a basic blank column chart. Click the *Select Data* button and then click *Add*. Select a cell for the label (the one containing the *Median* heading), and then select the range of cells containing the median values. Click *OK* to return to the *Select Data Source* dialog.

14. Now click the *Edit* button to alter the category labels. Select the six rows and two columns containing the labels for the *Wool* and *Tension* variables but not the heading labels. Click *OK* twice to return to the chart.

15. Click the *Layout > Error Bars* button and select the *More Error Bars Options* option. Click the *Custom* radio button then *Specify Value*.

16. You may have to click *Cancel* and then move the previous dialog out of the way. If so then return and then you'll be able to select the cells in the *Up* and *Dn* columns. In the *Positive Error Value* section select the cells in the *Up* column (only the values, not the heading label). Then in the *Negative Error Value* box select the cells in the *Dn* column. Click *OK* then *Close* to return to the chart.

Your final chart should resemble Figure 8.20. You may need to do some editing to get it to that point, by deleting the legend and title for a start.

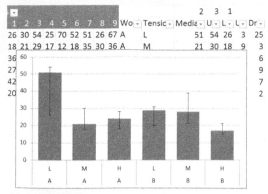

Figure 8.20 A bar chart with non-parametric error bars.

Once you have the basic chart you can use the filters to display certain portions of the dataset. You may have to move the chart to a new spot so that it doesn't get truncated when the filter hides some rows.

In the preceding exercise you had the option of copying the non-parametric summary data to a new worksheet. If you chose not to, but to keep the data in place then you'll have noticed that you were able to filter the data by selecting the block explicitly. The advantage of keeping the data in place is that you can rearrange the order of the samples in the pivot table. However, if the pivot table is altered in any other way, your summary data will be scrambled.

If you have a lot of replicates you may choose to arrange your pivot table in sample columns, rather than rows. This is not a problem as you can construct your chart just as easily with the values in that orientation. However, you won't be able to use the filter tools.

Tip: Rotating summary data
If you make your summary data table with rows for each statistic, rather than columns, you can simply rotate the table using *Paste Special* and the *Transpose* option.

Bar charts are the most widely used type of graph for displaying differences data but there is an alternative, the box-and-whisker plot, which you'll see next.

8.2.4   Box-and-whisker plots

A box-and-whisker plot (also called simply a boxplot) is a graph that conveys a lot of information in a small space. In general a box-and-whisker plot displays the median values for data in sample groups. The variability is shown in two ways; the box part shows the inter-quartile range and the whiskers show the max–min, i.e. the range (Figure 8.21).

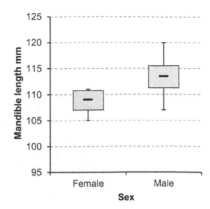

Figure 8.21 A box-and-whisker plot conveys a lot of information. Mandible lengths for golden jackal (mm), points = median, boxes = inter-quartile range, whiskers = range.

In Figure 8.21 you can see the mandible lengths for two samples of golden jackal, males and females. The box-and-whisker plot enables you to see how the samples relate to one another as well as an impression of the general shape of the data distribution. The *Female* sample for example appears to be a little skewed, whilst the *Male* sample appears more symmetrical and normally distributed.

Although box-and-whisker plots use non-parametric data they are commonly used for data that are normally distributed. If a sample is normally distributed the mean and median coincide and the inter-quartile range (IQR), whilst not directly analogous to standard error, gives a good impression of variability.

Excel is able to produce quite effective box-and-whisker plots (Figure 8.21 was produced using Excel 2010) but you need to use a little persuasion to produce the chart you require, as you'll see next.

Preparing the data

Before you can make a box-and-whisker plot you'll need to prepare the appropriate summary statistics. This means you'll need to arrange your data into sample groups using a pivot table. There is no box-and-whisker plot button in Excel and you'll have to press into service a different sort of chart, a *Stock Chart*. The type you need is called an *Open-High-Low-Close* chart and requires four values (as the name suggests). The four values correspond to the summary statistics as follows:

- Open – forms one edge of the box; you require one of the inter-quartiles.
- High – forms a point (usually not displayed); you require the median.

- Low – forms a point (usually not displayed); you require the median.
- Close – forms one edge of the box; you require one of the inter-quartiles.

Essentially you use the *Open* and *Close* values to form the box part. You use the *High* and *Low* values to make a point by displaying only one of the two values. The whiskers are formed by using error bars. However, the error bars require a deflection so you'll have to work out the difference between the median and the max and min values. You did something similar in the exercise in the preceding section, when you built a bar chart with error bars.

The data for the jackal box-and-whisker plot (Figure 8.21) were prepared and laid out in the spreadsheet as shown in Table 8.1.

Table 8.1 Data prepared for assembly as a box-and-whisker plot need to be in a particular layout. The first four columns are used for the main chart (boxes and median points). The last four columns are used to make the error bars (the whiskers). Each column forms a sample group.

Sex	UQ	Median	Median	LQ	Max	Min	Up	Dn
Female	110.75	109	109	107	111	105	2	4
Male	115.5	113.5	113.5	111.25	120	107	6.5	6.5

You can see from Table 8.1 that you need to calculate all the quartiles (the median twice). You'll use the inter-quartiles and the medians to make the main boxes, and the max and min values to construct the error bars (the deflections from the median to the extremes).

If your data contain multiple grouping variables you simply construct your pivot table so that the samples are in the groupings you want to display (placed in columns on the far left of the table). It would be slightly easier to set out your data so that the samples were in columns and the data series (the summary statistics) were in rows. However, in the orientation shown in Table 8.1 you can use filter tools to manipulate the displayed data.

If you find it easier to lay out your summary data in the other orientation (that is, with samples as columns and statistics as rows) you can use *Copy* and *Paste Special* with the *Transpose* option set. This rotates the data to the other layout.

In summary, here are the steps you need to take to prepare your data for a box-and-whisker plot:

1. Make a pivot table; you'll need index variables so that you can set out your data in sample groups. Have the predictor variables (the groupings) as the *Row Labels*, have the index variable as the *Column Labels* and place the response variable in the *Values* section.
2. To the right of the pivot table repeat the headings for the grouping variables and their labels. Then make headings for the summary statistics (see Table 8.1).
3. Use the QUARTILE function to help you calculate the summary statistics. Then work out the values for the error bars, which are the absolute difference between the extreme values (max and min) and the median.

4. You can either use the data in place or *Copy* and *Paste Special* to a new worksheet. You need to keep the values and number formatting. Select the predictor variables (including their headings) and then click *Data > Filter* to set up filters for these variables.

Once you have your data prepared you can move on to create the box-and-whisker plot. In the following exercise you can have a go at preparing some data; you'll use these data to make a box-and-whisker plot later.

Have a Go: Prepare non-parametric summary data for a box-and-whisker plot
You will need the *jackal.xlsx* file for this exercise. The data show the lengths of jawbones for golden jackals. The spreadsheet contains two worksheets; *Sample-layout* and *Recording-layout*. The *Recording-layout* worksheet, the one you'll use, shows the *Length* of the mandible (in mm) and the *Sex* of the individual. The *Observation* column is a simple index variable.

↘ Go to the website for support material.

1. Open the spreadsheet and navigate to the *Recording-layout* worksheet. This contains the data in the usual scientific recording format. Start by making a pivot table: click once anywhere in the block of data then use the *Insert >PivotTable* button and choose to place the result in a new worksheet.
2. Build the pivot table by dragging the fields from the list at the top to the sections at the bottom: *Length* to *Values*, *Sex* to *Row Labels* and *Observation* to *Column Labels*.
3. Now go to the *Design* menu and turn off all totals using the *Grand Totals* and *Subtotals* buttons. Click on the *Report Layout* button and set to *Show in Tabular Form*. You should now have two rows, one for *Female* and one for *Male* samples.
4. Click in cell L4 and type a formula to make a label for the grouping variable, =A4. Now copy this down the column to make labels for each of the two samples, so you want to end up with labels in cells L4:L6.
5. Click in cell M4 and type a label for the first of the summary statistics, UQ (the upper quartile). Move across the row and type more headings: Median, Median, LQ. In the cells above the labels (row 3) type the following values: 3, 2, 2, 1. These values correspond to the quartiles that you will calculate shortly.
6. Click in cell R4 (note that you will leave column Q blank). Now type a label for the maximum, Max. Continue across the row and type more labels: Min, Up, Dn. Now in cell R3 type the value 4 and in S3 type 0 (zero); these relate to the quartiles represented by the max and min values.
7. Now go to cell M5 and type a formula to calculate the upper (third) quartile, =QUARTILE($B5:$K5, M$3). Note that you must fix the column references for the first part using the $. The M3 part points to the value 3;

fix the row with the $. You can use the mouse to select the cells then edit the formula afterwards.

8. Copy the formula in cell M5 and paste it across the row to fill up to column S; do not worry about the blank column Q. Now copy the row down so that the summary statistics are calculated for both samples.

9. Highlight the cells in column Q and press the delete key on the keyboard to remove the values, restoring the blank column.

10. Click in cell T5 and type a formula to calculate the error bar, that is the difference between the median and the maximum: =ABS($N5-R5). Note that you need to add a $ to fix the column containing the median. Copy the cell across to the next column and then down to fill the next row.

11. Use the mouse to highlight cells L4:L6, then click the *Data > Filter* button.

You have now produced non-parametric summary statistics that can be used to make a box-and-whisker plot. Your data should look much like those in Table 8.1. You've also prepared the filter for the grouping variable (predictor variable), which will allow you to display portions of the dataset. In this case there are only two groups so this is hardly necessary. However, it illustrates that it is possible to make a data filter without having your data in a new worksheet.

The data that you prepared in the preceding exercise will be used to make a box-and-whisker plot shortly.

Building a box-and-whisker plot

In order to make a box-and-whisker plot you need to use an appropriate template; there is no specific chart type for a box-and-whisker plot in Excel but there are *Other Charts* available. One such chart is in the *Stock* section (Figure 8.22).

Figure 8.22 You can make a box plot using one of the *Stock* charts, available via the *Other Charts* button on the Insert menu.

The chart is called an *Open-High-Low-Close* chart; not a very snappy title but it does sum up what it does. The chart requires four values, which you've already calculated. In order to make the box-and-whisker plot you have to be somewhat creative, which is why the median value is calculated twice, as you'll see shortly.

It is not possible to build this kind of chart starting from a blank (that is, empty) starting point. You have to highlight the appropriate data first; this has some quirks, as you'll see from the following general procedure:

1. Start by preparing your data. You'll need four sets of summary statistics, the upper quartile, the median, the median again and the lower quartile. The max and min values are not required directly; you need them solely to calculate the error bars, which are added after the chart is made.
2. Your data are best set out with the summary statistics in columns. This allows you to have your grouping variables in columns, which you can then filter. The stock chart makes a chart with the statistics in rows, but you can't use filters.
3. Highlight the data that form the summary statistics, including their label headings. This is where the quirkiness of Excel comes to the surface. You have to highlight the four columns of summary statistics of course. You also have to have highlighted at least four rows of values (that is five rows including the labels). However, if you have exactly four rows highlighted the chart will be produced the wrong way around! So, you must highlight four columns and at least six rows, even if they are blank (Figure 8.23).

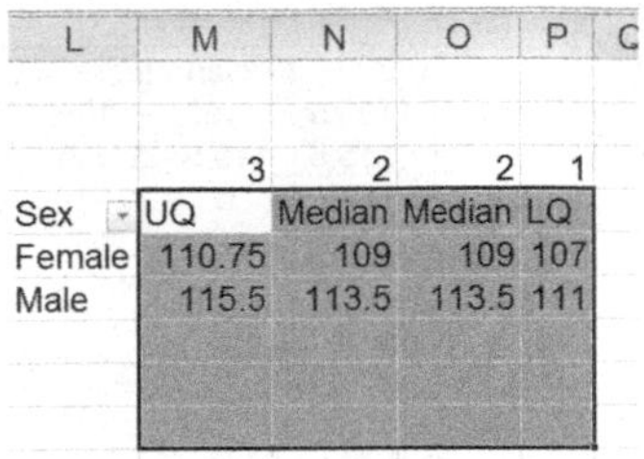

L	M	N	O	P C
	3	2	2	1
Sex ▾	UQ	Median	Median	LQ
Female	110.75	109	109	107
Male	115.5	113.5	113.5	111

Figure 8.23 You need to highlight at least six rows of data, even if blank, in order to get the *Stock* chart to produce a result.

4. Once you've highlighted the correct number of columns and rows you can click the *Insert > Other Charts* button. Then select the *Stock* chart labelled *Open-High-Low-Close*.
5. You now get a chart but there are probably additional blank data series (Figure 8.24). The data series themselves will not be labelled; it is easier to add the labels separately. You'll notice the selection borders in the worksheet if you click on the chart. You could use these to re-define the data but use the *Select Data* button instead.
6. Use the *Select Data* button to open the *Select Data Source* dialog box. In the *Chart data range* box you'll see the currently selected data. Delete this and click on the cell containing the first label for your grouping variable. Select all the data; include the grouping variables and the summary statistics, with their headings (Figure 8.25). If you have lots of rows it is easier to use keyboard shortcuts (Ctrl+Shift+Arrow). Click *OK* once you are done and your chart should now display the correct samples and their names.

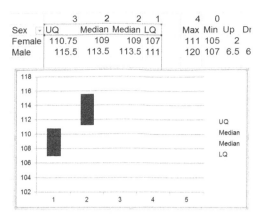

Figure 8.24 The newly created box-and-whisker plot contains no grouping labels and usually has additional blank series.

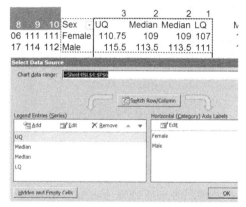

Figure 8.25 The *Select Data* button allows you to select the appropriate data for the box-and-whisker plot. Including the grouping variable label allows the axis labels to be formed correctly.

7. You should now have a basic chart. Most likely all you'll see are black rectangles, which represent the inter-quartile ranges of the various sample groups. Click on the chart to activate the *Chart Tools* menus. Then go to the *Layout* menu. On the left you can select the various chart elements; this is a good deal easier than trying to click on the items directly in the chart (Figure 8.26).

Figure 8.26 The *Layout* menu allows you to select chart elements from a list.

8. Select the item *Down-Bars 1* then click the *Format Selection* button. Set the fill colour of the boxes to something lighter: from the *Fill* section choose *Solid fill* and then pick a medium grey. Click *Close* once you are done and now the boxes appear in your selected colour.

9. Now select the item *Series "Median"* from the list in the *Layout* menu. There are two items; pick the top one. You want to make a marker appear so click *Format Selection* then go to the *Marker Options* section. You'll see that the *None* option is selected. You can use an *Automatic* marker or choose one yourself; click the *Built-in* option then change to a type you prefer (I like a kind of dash). Then change the *Marker Fill*; make it *Solid fill* and choose a dark colour (black is simplest and clearest). Now go to the *Marker Line Color* section and set that to *Solid line* (or *None*). Click *Close* when you're done; your chart now has median markers and inter-quartile boxes (Figure 8.27).

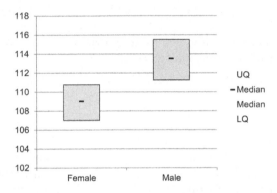

Figure 8.27 Once the down-bars are formatted and the median markers are added the box-and-whisker plot looks useful.

10. Now you can add the error bars. Click on the chart then go to the *Layout* menu. Click the *Error Bars* button and choose *More Error Bars Options*. You are asked which series you want to add the bars for (Figure 8.28).

Figure 8.28 With multiple series you can choose which to add error bars to.

11. Choose the top *Median* entry and click *OK*, which will open the *Format Error Bars* dialog box. You'll need to be able to see the cells containing the values for the error bars so you might want to move the dialog box. Click the *Custom* option then the *Specify Value* button to open the *Custom Error Bars* dialog box (Figure 8.29).

Figure 8.29 You can specify exact values for positive and negative error bars.

12. Use the mouse to select the appropriate values for your error bars. Then click *OK* to return to the *Format Error Bars* dialog box. You can apply other formatting as you like (for example, make the bars wider). Click *Close* when you are done to return to the chart.
13. Apply any other formatting you need to make your chart, for example you can delete the legend and make the gridlines dashed.

The process is slightly involved but once you have the hang of it things are relatively straightforward.

Tip: Save a box-and-whisker plot template
Once you've made one box-and-whisker plot, you can save the chart as a template. Click your chart then click the *Design > Save As Template* button. You can then apply the basic formatting to any data once you've made the basic chart and reselected the correct data. You'll have to add error bars independently but the large bulk of the formatting will have been done for you.

In the last exercise you prepared some data for use in a box-and-whisker plot. In the following exercise you can use these data to make a box-and-whisker plot.

Have a Go: Make a box-and-whisker plot from non-parametric summary data
You should already have the appropriate summary statistics set out for this exercise. If you haven't then look back to the preceding exercise and follow the steps to make the appropriate summary statistics for the *jackal.xlsx* dataset.

↘ Go to the website for support material.

1. Navigate to the worksheet that contains your pivot table and the summary statistics you prepared earlier.
2. Click in cell M4 then highlight the cells across to column P (the four summary statistics) and also drag the mouse to highlight several rows; you want to select cells M4:P9. You should end up with a block of cells four columns wide and six rows deep.

3. Now click the *Insert > Other Charts* button and select the *Open-High-Low-Close* chart in the *Stock* section. You'll now have a basic chart.
4. Click the *Select Data* button. Delete anything in the *Chart data* range box. Click in cell L4 to set the top-left of the data selection, now use Ctrl+Shift+Right-Arrow on the keyboard; this will select the heading row (this is why you left column Q blank). Next, press Ctrl+Shift+Down-Arrow on the keyboard. This will select all the rows of data. Now click *OK* to return to the chart.
5. Click the *Layout* tab and select the *Down-Bars 1* item from the dropdown list on the left. Then click *Format Selection* and alter the fill colour of the boxes to a medium grey. Click *Close* when you are done.
6. Now select the *Series "Median"* item (the first of two) and format that to show a marker in an appropriate colour.
7. Click the *Layout > Error Bars* button and select *More Error Bars Options*. Now choose *Median* as the series to use for the error bars.
8. The *Format Error Bars* dialog box will be open. Select *Custom* then click the *Specify Value* button. In the *Positive Error Value* box select the cells T5:T6, which are the ones under the *Up* heading in the worksheet. In the *Negative Error Value* box select the cells U5:U6, which are under the *Dn* heading. Click *OK* then click *Close* to return to the chart.

Your chart is now complete but you can edit it to tidy it up somewhat. The legend is not really needed and the gridlines are too strong. Do some editing using the *Chart Tools* menus; your final chart should resemble Figure 8.21.

The box-and-whisker plot you made in the preceding exercise is quite simple; there is only one grouping (predictor) variable and this only contains two samples. However, the processes involved are the same, no matter how complicated your dataset is. As long as you set out your pivot table with the grouping variables in the *Row Labels* section and have the appropriate index variable you can follow the same process.

Note: Box plot template
A chart template that contains much of the appropriate formatting is available as part of the example data to accompany this book. The file is *boxplot.ctrx*. You need to add it to your chart template folder for it to be available. Click the *Change Chart Type* button, then *Manage Templates*. You'll then be able to drop the template file into the correct folder/directory.

↘ Go to the website for support material.

Once you have an appropriate pivot table, the mechanism for creating the summary statistics and subsequent box-and-whisker plot is the same however complex your dataset is. You can use filters to alter the groupings that you display, thus you are able to visualize logical portions of your dataset quite easily (Figure 8.30).

		3	2	2	1	4	0		
Woi ▾	Tensic ▾	UQ	Median	Median	LQ	Max	Min	Up	Dn
A	L	54	51	51	26	70	25	19	26
A	M	30	21	21	18	36	12	15	9
A	H	28	24	24	18	43	10	19	14
B	L	31	29	29	20	44	14	15	15
B	M	39	28	28	21	42	16	14	12
B	H	21	17	17	15	28	13	11	4

Figure 8.30 These summary data contain two grouping variables. When set out in columns the *Filter* tool can be used with a box-and-whisker plot to visualize portions of the dataset.

If you have multiple grouping variables they appear on multiple lines of the *x*-axis (Figure 8.31).

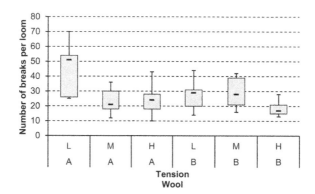

Figure 8.31 A box-and-whisker plot using multiple grouping (predictor) values. If the data are filtered you can view a subset of the original dataset.

Earlier, when you looked at bar charts, you used a dataset with two grouping variables (*warpbreaks.xlsx*). Try this out as an additional exercise. You should be able to set out your pivot table and subsequent summary statistics using the guidelines you've seen and produce a result similar to Figure 8.31.

8.3 EXERCISES

1. Which of the following summary statistics **cannot** be calculated directly in a pivot table?

   A) Sum
   B) Mean
   C) Standard error
   D) Variance
   E) Replicates

2. Which of the following types of chart can be used for visualizing differences data (choose all that apply)?

   A) Bar chart

B) Dot chart
C) Line plot
D) Scatter plot
E) Pie chart
F) Box-and-whisker plot

3. You can use the upper and lower quartiles (which form the IQR) as values in error bars. TRUE or FALSE?

4. Box-and-whisker plots usually show ____ statistics for example, the ____, ____ and ____.

5. Look at the *Butterflies and Site Management.xlsx* data file, which gives a measure of the abundance of different butterfly species measured in different years at a nature reserve. The reserve was split into four transects (corridors along which the butterflies were counted). Each transect has several sections; these have numbers and also a management designation. Some are managed in a particular way and others not. Look at data from 1997 and focus on the *swf* species. What are the medians and IQR for the sections with and without management? You can think of the *Section* variable as being an index variable.

↘ Go to the website for support material.

The answers to these exercises can be found in Appendix 1.

8.4 SUMMARY

Topic	Key points
Differences data	With this kind of dataset you are looking for differences between sampling groups. Your main aim is to split the data into meaningful chunks and to compare them.
Summarizing sample groups	You can use filter tools and pivot tables to help you arrange your data by sample group.
	The aims are to look at measures of centrality (average), dispersion (variability) and replication.
Pivot table summaries	Pivot tables can use certain summary statistics: sum, count, average (mean), max, min, standard deviation and variance. You cannot use any non-parametric statistics.
	You can alter the summary statistic via the *Value Field Settings* option.
	Items in the pivot table can be reordered (the default is alphabetical).
	You can use a field multiple times in the *Values* section, and alter the summary statistic associated with each field separately.
Non-parametric statistics	Pivot tables cannot do non-parametric statistics so you must either make a filter or use index variables and a pivot table, then use regular Excel functions as required (for example, MEDIAN).
	It can be helpful to copy and paste a pivot table using *Paste Special* to copy only the values, thus breaking the link to the pivot table and fixing the data.

Topic	Key points
Calculated fields	In a pivot table you can use a calculated field to summarize using another function, for example, *Log10*. You enter a regular formula in the calculated field box but any functions must operate on single values (that is, not ranges of cells).
Analysis ToolPak	The *Analysis ToolPak* cannot handle grouping variables. However, you can use data in a pivot table. So to use groupings make a pivot table first, then use the *Descriptive Statistics* option from the *Analysis ToolPak*.
Visualizing differences data	There are two main sorts of graph to help visualize differences data: bar charts and box-and-whisker plots.
Sparklines	Sparklines can be produced as a mini in-cell bar chart from the *Insert* menu to act as a quick overview. There are limited formatting options but you can highlight the largest and smallest bars.
	Sparklines can contain regular cell contents in addition to the chart.
Histograms	Histograms help you to explore data distribution. You need to make a pivot table first, to arrange your data in sample groups. You can then use the *Analysis ToolPak* and the *Histogram* option.
Bar charts	Pivot charts can be used to make bar charts. These can only show parametric data (for example, the mean) and it is not easy to add error bars to show variability.
	A bar chart that you build can use any statistic and error bars are easily incorporated.
	If your data are arranged appropriately you can use *Filter* tools and alter the chart accordingly.
	Usually you place all the predictor variables in the first columns (the *Row Labels* area of a pivot table).
	If your bars vary a lot in height you can use a logarithmic axis to rescale them.
Error bars	Error bars often display the standard deviation or standard error. Non-parametric error bars show the range or IQR. Note that error bars show the deviation from the middle value so standard error works as it is but IQR needs further work (to compute differences from the median).
	You can add error bars to any chart but it is easier if your chart is not a pivot chart.
Box-and-whisker plots	Box-and-whisker plots display a lot of information in a compact space. Usually you display the median, IQR and max–min values.
	You need to compute the various statistics and can use a pivot table as the starting point.
	You cannot build a box-and-whisker plot from a blank chart so you'll need to select cells first then edit the data afterwards.
	Insert a box-and-whisker plot from the *Other Charts* section of the *Insert* menu. Use an *Open-High-Low-Close* type of *Stock* chart.
	If your data are arranged appropriately you can use *Filter* tools and alter the chart accordingly.
	If you make a box-and-whisker plot you can save it as a chart template, which can save some formatting steps another time.

9

SHARING YOUR DATA

Sharing your data is an important part of the scientific process. In a general way this is how scientific knowledge is gained and our understanding of the world advances. Ideas are tested by collecting data and exploring the results. These results are then shared with the scientific community and the science advances.

In a practical sense there are various potential recipients you can share your data with:

- Colleagues in your workplace, school or research group.
- Scientists in your discipline.
- Your supervisor: team-leader, teacher, or professor for example.
- A client or grant-awarding body.
- The general public.
- Computer software.

Exactly what you share depends on who you are sharing with. If you are working in a research team you may share more or less everything. They are probably familiar with what you are up to so some details might be omitted. Other scientists in your discipline might not be so familiar with what you are doing so you might share your data with them in a slightly different manner. You'll probably stick to sharing your results, rather than all your data.

You may have to report to a supervisor; they'll probably want to see an overview to check progress. A client or grant-awarding body will want to see what their money was spent on! Whilst the general public might want to know how their tax was spent, in general they will want to see results but not the minutiae.

Excel is a powerful piece of software but it is generally more suited to managing your data than carrying out detailed statistical analyses. Excel can conduct a range of summary statistics, as you've seen, so it is good for data exploration. You can also use Excel to produce some useful graphs, although you often need to bully the software to produce the chart you want.

Although Excel can carry out some statistical tests it was never designed for such work, and its repertoire is limited. For example: you can carry out a t-test, which will examine the difference between two normally distributed samples, but you cannot undertake the same task for skewed data, as the U-test is not available in Excel (although you may be able to get a 3rd party add-in). Excel is quite good at regression analysis, as long as your data are normally distributed, but most other kinds of analysis are only available to Excel if you carry out the maths yourself or get 3rd party add-ins.

So, in the main you'll probably need to share your data with another program, more suited to statistical analysis. There are many statistical programs available; some are free and others are not.

You can share three elements of your data:

- The data itself.
- Numerical summaries of your data.
- Visual summaries of your data.

If you carry out statistical analyses of your data obviously you'll want to share those too. However, since this book is about the early stages of the scientific process you'll see only the basics, under the three main groups as detailed in the preceding list.

9.1 EXPORTING YOUR DATA

There are many reasons why you may need to share your dataset. What you share can be broken down into two elements:

- The entire dataset.
- Some part (or parts) of the dataset.

If you have an Excel file then it seems obvious that you can simply use that. However, your file may contain additional information that you don't want (or need) to share. Also, the Excel file format may not be suitable; this is especially the case if you need to carry out further statistical analysis in a different computer program.

9.1.1 Sharing your entire dataset

If you've followed the protocols set out in the early chapters of this book, your Excel file will probably contain a main worksheet with the actual data and other worksheets. These other worksheets might contain:

- Notes and methodology.
- Pivot tables.
- Charts.
- Lookup tables.

If you want to incorporate everything then you can simply send the entire file. The default format since Excel 2007 is XLSX. This format allows elements to be incorporated into spreadsheets that were formerly not possible. Generally these are related to graphics such as *SmartArt*, which can be found under the *Insert* menu (Figure 9.1).

If you are sharing a file with someone with an old version of Excel some elements of your spreadsheet may not be seen properly (or at all). In the main it is best to stick to

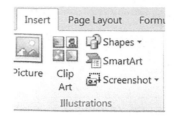

Figure 9.1 Some items, such as *SmartArt*, contain XML and cannot be saved in older XLS Excel formats.

simple spreadsheet elements, which are less likely to appear wrongly when shared. Your aim should be to provide the maximum shareability for your dataset.

Save in a different format

If you need to save your file in a different format than the default XLSX then you'll need to use the *File > Save As* button. You can then select a location for the saved file and the type; use the dropdown icon in the *Save as type* box to bring up an extensive list (Figure 9.2).

Figure 9.2 Use the *File > Save As* button to choose from an extensive range of file types.

In general you'll only need to use one of two types:

*   Excel 97–2003 Workbook.
*   CSV (comma delimited).

The first option allows you to save the entire spreadsheet in the older XLS format, which is readable by older versions of Excel. The second option allows you to save only a single worksheet (the currently active one). The CSV format (comma separated values) is readable by more or less every data analysis program. This makes CSV a good choice if you need to export your data to a specialist analytical program. Any formatting is removed from CSV files and all that remains are the data. You can open CSV files in many programs, including word processors and text editors (Figure 9.3).

When you open a CSV file in Excel it reads each line as a new row. The columns are defined by commas. So, each time a comma is encountered Excel starts a new column.

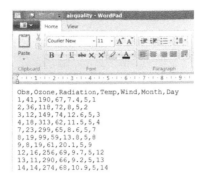

Obs,Ozone,Radiation,Temp,Wind,Month,Day
1,41,190,67,7.4,5,1
2,36,118,72,8,5,2
3,12,149,74,12.6,5,3
4,18,313,62,11.5,5,4
7,23,299,65,8.6,5,7
8,19,99,59,13.8,5,8
9,8,19,61,20.1,5,9
12,16,256,69,9.7,5,12
13,11,290,66,9.2,5,13
14,14,274,68,10.9,5,14

Figure 9.3 A CSV file can be opened in a text editor. The columns are separated by commas. New rows are indicated by new lines (the end-of-line character is not visible).

Note: Numbers in CSV files
When you save a worksheet as a CSV file the number format that was in operation becomes fixed. So if you used two decimal places to view the data then this is all you end up with in the CSV file. If you want to keep all the decimals you need to ensure the correct format is in operation before you save the file in CSV format.

Tip: After saving as CSV
When you save a file as CSV Excel gives you warnings about multiple sheets and formatting. Read the messages carefully but remember that you are making a copy of the file on disk. The saved file will contain a single worksheet but you'll still see all the worksheets visible in the copy that is open on screen. It is a good idea to close the spreadsheet and reopen the original Excel format file, especially if you intend to do anything else to the file.

The CSV format is very useful because:

- You can share the data with any computer program.
- You strip out unnecessary formatting and formulae, which reduces file size.

There are limitations; you can only handle one worksheet at a time. There are times when you want to save a single worksheet to disk in regular format; this is what you'll see next.

Save selected worksheets
Your spreadsheet may contain multiple worksheets but you might only want to share one of them. There are two options:

- Save a copy of the entire spreadsheet then delete the worksheets not required.
- Copy a single worksheet directly to a new workbook.

The first option obviously allows you to keep one or more worksheets, which could be useful if you wanted to share your main dataset and the notes for example. You can delete a worksheet in several ways:

- Right-click the worksheet name tab and select *Delete*.
- Use the *Home > Delete > Delete Sheet* button.

If you right-click on a name Tab for a worksheet you open a new menu with various options (Figure 9.4).

Figure 9.4 Right-clicking a worksheet name tab brings up a menu of options allowing you to move or copy a worksheet.

You can make a copy of the worksheet directly in a new spreadsheet (workbook) like so:

1. Right-click the name tab of the worksheet you want to make a copy of. This opens a pop-up menu (Figure 9.4).
2. Choose the *Move or Copy* option, which opens a new dialog box (Figure 9.5).

Figure 9.5 The *Move or Copy* dialog allows you to move a worksheet or make a copy. The target can be a new spreadsheet (workbook).

3. Click the dropdown icon in the section labelled *Move selected sheets to book:* then choose a new workbook. All open workbooks are listed as well as the option *(new book)*.
4. Make sure you tick the box labelled *Create a copy*, otherwise the worksheet will be moved rather than copied. Click *OK* to complete the operation.

5. Edit the new workbook as you like and save to disk when you are ready. You can of course save the result in a format of your choosing.

It is a good idea to rename the worksheets to something appropriate; short and descriptive is best. You can rename a worksheet by right-clicking the name tab and selecting *Rename* or by using the *Home > Format > Rename Sheet* button.

> **Tip: Select multiple worksheets**
> You can select more than one worksheet by clicking on their name tabs. Hold the Ctrl key to select multiple sheets that are non-adjacent. The Shift key allows you to select a block of tabs. Once selected you can right-click on any of the tabs to open the dialog allowing you to move or copy several sheets at once.

9.1.2 Sharing portions of your dataset

You may not want to share the entire dataset. The scientific recording format that you should be using allows you to use your data as if it were in a database. This means you can select records matching certain criteria quite easily.

If you use a filter you can display only those records matching one or more criteria. The problem is that the regular filter merely hides the data that don't match; even if you save as CSV you'll find that all the data are saved, unhidden. You need to use the *Advanced* filter, which you met earlier (Section 2.2.3).

The *Advanced* filter allows you to match several criteria and send the results directly to a new workbook. You can then save the workbook as a regular Excel file or as a CSV, as you like. Here is a reminder of the steps you need:

1. Make a new workbook (spreadsheet file) to act as the recipient of the filtered data.
2. In the new workbook make sure you have at least two worksheets. One will hold the filter criteria, the other the final filtered data.
3. Open the source dataset and navigate to the worksheet containing the data you wish to filter.
4. Select the top row of the data, containing the headings, and copy to the clipboard.
5. Navigate to the new workbook and paste the header row into one of the worksheets.
6. In the rows under the headings enter the criteria you want to use.
7. Go to the blank worksheet where you want the results to end up.
8. Click the *Data > Advanced* button to open the *Advanced Filter* dialog box.
9. Click the radio button next to the option *Copy to another location*.
10. Click in the *List range* box and select the original data (the source data). You can click in any cell then use Ctrl+A to select the entire dataset.
11. Now click in the *Criteria range* box and select the cells containing the filter criteria (from steps 3–6).

12. Click in the *Copy to* box. Navigate to the target worksheet and click in cell A1.
13. Now click *OK* to complete the filter and place the results in the new worksheet.
14. Save the new workbook. You can either keep the filter criteria or not. If you save the filtered records as a CSV file the worksheet with the criteria is not saved.

If you are intending to filter out records more than once you might consider adding a new worksheet containing the header row to the main dataset. This will save you a step each time, as you'll only have to edit the criteria rather than copying and pasting the header cells.

9.2   SHARING YOUR SUMMARIES

You won't always be dealing with the original data. You've already seen how to summarize data in more concise ways, using averages and so on (Chapters 4–8). These numerical summaries are more easily digestible than the original data. You may need to share these summaries in order to give an overview to your colleagues, supervisor or the general public.

Your summaries should be in two main forms:

* Pivot tables.
* Regular spreadsheet cells.

The means by which you share these items will depend to some extent on your purpose. If you are giving a talk or presentation you'll probably want to copy the information to PowerPoint. If you are creating a printed report you'll need to transfer the data to your word processor. At other times you may simply want to share the spreadsheet data in a more informal manner and Excel itself may be an appropriate target.

You can think of your target media as being in two main groups:

* Other Office programs.
* Non-Office programs.

In general the Office suite shares data between programs easily and allows you a good degree of control over the results. At other times it may be more appropriate to use CSV format or something else (such as PDF).

9.2.1   Sharing pivot tables

Pivot tables are powerful and flexible; you can alter one quickly and easily. There are two main routes you can use to share a pivot table:

* Using *Copy* and *Paste*.
* Saving as a file of some kind (for example, CSV, XLSX, PDF).

Use copy and paste
If you copy a pivot table to the clipboard you can transfer it to another Office program

and retain the ability to alter the table. Essentially you can embed the underlying spreadsheet into the target program. There are benefits and downsides to this behaviour. The main benefit is that you can alter the result, giving a measure of flexibility. The downsides are:

- You may not want the recipient to be able to alter your pivot table.
- The embedded spreadsheet can make file sizes very large.
- Sometimes the target file contains a link to the original spreadsheet, which may be missing when you send the file to the recipient.

The key to controlling the result is in using the *Paste* button and selecting the most appropriate format for the layout. The two most commonly used Office programs you'll use are Word and PowerPoint. In Table 9.1 you can see the range of *Paste* options available for Word.

Table 9.1 *Paste* options for Microsoft Word.

Option	Result
Keep Source Formatting (K)	This makes a table object that retains the formatting of the original.
Use Destination Styles (S)	This makes a table object without any formatting. Essentially a bare Word table.
Link & Keep Source Formatting (F)	This makes a table with the original formatting but the table is linked to the original spreadsheet. The pivot table can be edited (right-click) but only if the file can be found (it is not embedded).
Link & Use Destination Styles (L)	This is similar to above except that the table retains no formatting.
Picture (U)	This makes an image of the table, stored in a picture format. The result is obviously not editable.
Keep Text Only (T)	This makes a text box containing the original table without any formatting. Items are separated by tabs and so may not always line up appropriately.
Paste Special	This opens a dialog that allows you to select the target type. In general you should only need this if you want to embed the original spreadsheet in the Word file.

The *Paste* options for PowerPoint are subtly different (Table 9.2).

Table 9.2 *Paste* options for Microsoft PowerPoint.

Option	Result
Use Destination Styles (S)	This makes a table in PowerPoint using the default table style for PowerPoint.
Keep Source Formatting (F)	This makes a table but retains the formatting from Excel.
Embed (E)	This embeds the spreadsheet into PowerPoint, allowing the pivot table to be edited as if it were in Excel.

Continued

Continued

Picture (U)	This makes an image of the table, stored in a picture format. The result is obviously not editable.
Keep Text Only (T)	This makes a text box containing the original table without any formatting. Items are separated by tabs and so may not always line up appropriately.
Paste Special	This opens a dialog that allows you to select the target type. You should not need to use this option.

If you use the *Paste Special* option you can choose the target style from a list. In general choosing *Microsoft Excel Worksheet Object* results in the pivot table and entire spreadsheet being embedded. Other options generally do not embed or link to the original file.

> **Tip: Checking paste target items**
> Once you've used the *Paste* button to transfer data to your target program, check it is in the form you want before carrying on. Click the object in the target program and see what menus appear. Try right-clicking the object; if you see items relating to linked files or worksheets you'll know that you have an embedded or linked item, which may not be what you intended.

Other methods of sharing

You can use other options to save pivot tables:

- Save the worksheet as CSV.
- Paste to a new workbook with or without formatting, then save that.
- Save as PDF. Since Office 2010 you can save PDF files (in 2007 you need to download a Microsoft add-in). You can save a worksheet or selected cells.

In general a basic table format is most suitable for pivot tables, since they are indeed tables. However, as you've seen you do have a range of possibilities.

9.2.2 Sharing summary results

If you need to share any regular spreadsheet cells you can do all the same things as for pivot tables or worksheets. You can copy and paste to other programs, paying particular attention to the target formatting. You can also save the worksheets as you need.

9.3 SHARING YOUR GRAPHS

Summary tables are useful but visual summaries carry more impact and are generally understood more readily, especially by those who are not familiar with your work. You've seen many examples of graphs (Chapters 4–8) in relation to the exploration of various kinds of dataset.

When you are making graphs for yourself you probably won't make them as complete as you would when you are going to share the graph with someone else. The charts you make when you're exploring your data help you understand the patterns in the data so you don't need to annotate them as fully. Of course if you are saving the charts for future reference you'll need to remember what is what.

There are two elements to consider when sharing graphs:

- What components need to be added to the graph to make it understandable?
- What format should the graphic be in for sharing?

First of all you need to make sure that you include an appropriate title or caption, and that axes are labelled and so on. Secondly you need to work out how the finished chart should be shared; you may want to transfer the graph to PowerPoint or save it in a particular format as a standalone file.

You have seen two main sorts of graphic: sparklines and regular charts. Sparklines are less useful, because they don't contain many editable elements.

9.3.1 Sharing sparklines

Sparklines are useful for early investigations of your dataset, since they are quick to produce and give you a sense of the patterns in the data. However, they cannot be edited like regular charts. Sparklines can still be useful to share, for the same reasons you used them in the first place; they give a quick impression of the patterns in the data.

Preparing and editing Sparklines

Your first task is to prepare the sparklines and decide how you want to present them. The *Sparklines Tools > Design* menu contains tools to help you manage the appearance of your sparklines. You can also resize the cells containing the sparklines.

You can consider using the sparklines in two main ways:

- As standalone mini-graphics.
- Alongside a summary table (for example a pivot table).

In terms of preparation there is not a lot of difference between the two. If you are going to use sparklines as standalone graphics then you need to use some additional cells for labelling the elements (Figure 9.6).

Sum	Tension ▼		
Wool ▼ L		M	H
A	401	216	221 A ▦ __ __
B	254	259	169 B ▦ ▦ __

Figure 9.6 A pivot table with sparklines. The individual sparklines have labels in the adjacent cells, which are copied with the sparklines when they are shared.

In Figure 9.6 you can see a pivot table and some sparklines in the adjacent rows. The rows have been labelled (with the wool type) so that the sparklines can be shared, each with its own label. If you wanted to share the table and sparklines together then the additional labels would not be necessary.

As far as preparation goes there is little else you can do. Your next move is to transfer the sparklines, with or without the attendant table, to the appropriate target.

Exporting sparklines

Sparklines are a special kind of spreadsheet cell; they contain the mini-charts and any other regular spreadsheet data, text, numbers or a formula. If you try to paste sparklines into a word processor or PowerPoint you'll end up without the sparklines but with any underlying cell contents; most of the time you'll simply get blank cells. This means that you'll need to use the *Paste* options in a similar way to how you use them when sharing pivot tables (Section 9.2.1). Earlier you saw some of the *Paste* options for Word (Table 9.1) and PowerPoint (Table 9.2).

You've got three main options when sharing with your sparklines:

- Copy to another Office program.
- Save the entire spreadsheet or the worksheet containing the sparklines.
- Save the sparklines as a standalone graphic file.

Copying sparklines to another Office program

The sparklines need to be pasted using a graphic format; this means that when you click the *Paste* button you'll need to select *Picture (U)* as the output style. You can attempt to use *Paste Special* and set the format to be *Microsoft Excel Worksheet Object* (in PowerPoint you can use *Embed (E)* from the *Paste* button) but this can result in large file sizes and problems with linked files. In the main it is best to avoid embedding or linking files because the originals may not always remain with your target. So, you'll end up sharing your file with someone and they may not be able to read your data. Worse still is that they may be able to access and edit the original data; this is definitely something you want to avoid!

If you want to include labels (which you should) you simply select the cells along with the sparklines. If you intend to incorporate the summary table (probably a pivot table) with the sparklines you'll need a slightly different approach. You can simply paste the summary table and sparklines as a graphic; this results in an image of the table and sparklines. For most purposes this is quite adequate. If you want to be able to edit the table data you can transfer the table and sparklines in two phases like so:

1. Prepare your sparklines. Most likely you'll have a pivot table or some spreadsheet cells and the sparklines alongside.
2. Open the target program, most likely Word or PowerPoint.
3. Select the spreadsheet cells and the sparklines that you want to transfer.
4. Return to the target application and then click the *Paste* button. Select the style you want; most likely this will be *Keep Source Formatting (K)* or *Use Destination Styles (S)*, which will make a table.
5. Your table is editable but does not contain the sparklines. Your first task is to select the cells of the table in Word or PowerPoint, and then choose to merge them. You will incorporate a graphic of the sparklines in this space. You'll need to use the *Table Tools > Layout > Merge Cells* button.

6. Now return to Excel. Select the sparklines and copy them to the clipboard. Return to the target application. What you do next depends on which program you are using.

7. If you are using PowerPoint you must paste the sparklines in as a graphic (*Picture (U)* option). Then drag the image into the table. You'll need to resize to get things to line up.

8. If you are using Word you must use the *Paste* button and select *Picture (U)* as the format. Paste the sparklines to a blank line then drag the graphic into the table. Alternatively you can click in the table cell (the merged ones from step 5), then use *Paste > Paste Special* and select a picture format; *Picture (Enhanced Metafile)* is probably best.

You'll end up with a good representation of your summary table and sparklines; the method works particularly well in Word (Figure 9.7).

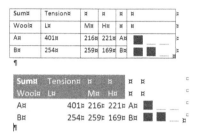

Figure 9.7 A pivot table and sparklines incorporated into a Word table. The upper table uses *Destination Styles* (S), whilst the lower uses *Keep Source Formatting* (K).

Saving sparklines in Excel

If you want to save a spreadsheet with a summary table and sparklines you can do this with a little help from the *Paste* button. Just follow these steps:

1. Prepare your sparklines; this may mean making a pivot table. In any event you need to arrange things how you want.

2. Once you are happy with the arrangement you need to make a new worksheet. Return to the sparklines and select the cells you want to transfer, probably your pivot table and the sparklines themselves. Copy them to the clipboard then navigate to the new worksheet.

3. Click the *Paste* button and choose the *Paste Values > Values & Source Formatting (E)* option. This strips out links to the original data but keeps the Sparklines.

4. Now right-click the name Tab and select *Move or Copy*, which will open a new dialog box.

5. Choose to move the worksheet to a new book, and then click *OK*.

6. You now have a fresh workbook containing a single worksheet. This should have your summary table and sparklines. You can simply save this in regular format (XLSX).

When you save your sparklines in this manner you lose any links to the original data. You can share the workbook or use it as the basis for sending to Word or PowerPoint, as you saw previously.

Saving sparklines as standalone graphics

There are times when you may find it expedient to save your sparklines as a separate entity in a graphical form. You have two main options:

- Save as a graphic file, for example, JPG, PNG.
- Save as a PDF.

In order to save as a graphic such as a JPG image you'll need to transfer the sparklines (and any table cells) to another program. PowerPoint can save slides as graphics so makes a good choice. Just use the methods outlined earlier to make a slide in PowerPoint, then use *File > Save As* to save in the format you want.

You can also open other programs and use paste operations to transfer the data. The *Paint* program comes with Windows and allows you to paste and then save in various graphics formats.

Tip: Saving quality graphics

If you are going to copy and paste a graphic from Excel to a standalone graphic you should make the source fairly large before pasting. This maintains the best quality. For best results copy your sparklines to an Office program (you can use Excel) then resize the graphic before copying to the clipboard.

If you want to save as a PDF you can do this directly from Excel. You simply select the cells you want then use the *File > Save As* button. Select PDF from the *Save as type* drop-down list. Click the *Options* button, you can then choose to save the selected cells or the entire worksheet.

Note: Office programs and PDF

In Office programs since 2010 you have the option of saving as a PDF. In older versions you can only save as PDF using an add-in. Check Microsoft's website for details of downloads for various Office versions.

9.3.2 Sharing charts

Using regular charts (as opposed to sparklines) is the most useful way to show summaries of your data to a wider audience. Regular charts can show more detail than sparklines and you can annotate and customize them more readily.

As you saw when dealing with sparklines, there are two phases to the sharing process:

- Preparing your charts.
- Transferring your charts.

It is worth spending a little time to prepare your charts carefully. These represent the climax of your research and are the ultimate method of presenting your work to others. If you produce a chart that looks substandard then others will think that your work itself is the same.

Preparing and editing charts

You've already seen how to produce basic charts to cover a range of data types (Chapters 4–8). Once you've got a basic chart you'll need to ensure that it contains certain elements before it is ready for sharing with the wider world. The elements you need to consider are:

- *Type* – Make sure you have the most appropriate type of chart for the data and the message you want to convey. In general simple graphs work better than complicated ones, so avoid the 3D charts.
- *Clutter* – Consider making more charts if one is too busy; the purpose of the chart is to aid understanding not add confusion.
- *Title* – You can incorporate an informative title on the chart itself or use a caption underneath. You can add a caption in PowerPoint or Word if you need to. Generally it is best to use a caption so that the chart area itself is devoted to the data, rather than a title.
- *Axis labels* – You should make sure that axes are labelled clearly so that the reader can see what is being represented. Don't forget to include the units of measurement. Any values should be formatted sensibly so if your data should be to two decimal places then set that format for the axis label.
- *Axis scale* – Try to rescale axes so that the data fill the plot area. It is not always appropriate to start axes from zero but do make sure that the axes are labelled clearly (see previous).
- *Legend* – If your chart includes data from more than one series you need to differentiate the series. A legend is the most sensible way to do this. You don't need a legend if there is only a single data series. If you do include a legend place it in what seems like the best place, don't just rely on the default setting.
- *Colour* – Colour is the simplest way to help readers differentiate between data series. However, consider your readers' requirements; red and green may not be a good combination if you have readers with a colour vision deficiency. The target may be a printed document, which will end up as monochrome. It is better for you to make a chart that is monochrome rather than hope that the colours you used can be differentiated later. You can use patterned fills for bars and different markers for points.
- *Points and markers* – Excel has a range of default markers. Sometimes these come out with shadows or 3D effects. These tend to obscure the patterns rather than help illuminate so consider using plain markers.

- *Fonts* – Some of the Microsoft Office fonts are not easily recognized by other programs. This is usually only a problem if you are intending to make your chart as a PDF. In the long run you are better off using standard Arial or Times New Roman, boring but safe. You are after legibility and portability.
- *Size and shape* – Excel charts have a generally rectangular shape. You can resize your chart to give the best effect, so if you have a column chart you may wish to have a tall chart rather than a wide one. A scatter plot might be better if it is nearly square.

Note: Excel charts and gridlines
Excel tends to add horizontal gridlines to all charts; these may not be appropriate, especially on scatter plots, so think carefully about their purpose and remove them if they aren't needed.

You can access all the tools you need to prepare your charts from the *Chart Tools* menus. You can select many chart elements by clicking on them directly; a right-click then allows you to edit the selected element. However, it can be fiddly so it is generally easier to use the selection window available from both *Layout* and *Format* menus (on the left in the *Current Selection* section). Once you have an item selected you can use the *Format Selection* button. The *Design* menu allows you to change chart type, select data and alter the chart colours. You can set general colours and fonts from the *Page Layout* menu, which is easier than formatting individual elements on the chart. If you prefer to format them separately then you can select an item and alter the font from the *Home* menu.

Tip: Using superscripts in axis titles
If you need to use superscripts in an axis title (for example, for m^2) then type your title as normal and then select the individual characters you need to change. Right-click and choose *Font*. You can then apply the superscript style to the selected characters.

Once you have your chart in a state that you consider acceptable to share you can move on to the next stage and transfer the chart as required.

Exporting charts
Once you've got your chart prepared you can share it in several ways:

- With Office programs, via copy and paste.
- As a graphics file, via another program.
- As a PDF file, from Excel.
- As an Excel worksheet.

Which option you choose depends on your purpose.

Copy charts to other Office programs

When you use the *Paste* button you have the same sort of options as when dealing with sparklines. You also have the same issues about embedding or linking to the source data. Embedding or linking allows you to edit and alter the chart from another Office program; however this is not always desirable. File sizes can be very large and you may not wish your recipient to be able to edit your data. In most cases using a picture format for the target paste option is your safest option.

The *Paste* button gives you the basic *Picture (U)* option but if you select *Paste Special* you get a wider choice (Figure 9.8).

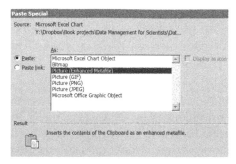

Figure 9.8 The *Paste Special* options in Word when dealing with a chart in the clipboard. The top and bottom options result in an editable chart (with embedded data). The other options are picture formats.

PowerPoint gives similar options. The *Microsoft Office Graphic Object* is essentially the same as the *Microsoft Excel Chart Object*.

Saving charts as standalone graphics

You can save a chart as a standalone graphics file in two ways:

- Paste to PowerPoint and use *File > Save As* to save the slide as a graphics file.
- Paste to another program, such as Paint, then save the file.

The first route is quite straightforward; you simply copy the chart then paste it into PowerPoint. Resize as you like then use the *File > Save As* button and select the format of your choice.

If you want to use something other than PowerPoint simply paste the chart into your chosen program. To get the best quality it is a good idea to make your chart quite large before copying to the clipboard.

Saving charts as PDF files

You can save your chart directly as a PDF from Excel. Select the chart (just click on it), then use the *File > Save As* button. Choose PDF from the dropdown in the *Save as type* box. Then click the *Options* button and make sure the option *Selected Chart* is checked.

It is a good idea to use standard fonts for your charts; some Microsoft Office fonts (Cambria and Calibri) are not always recognized properly by some programs that deal with PDF files (such as Adobe Illustrator). You can change the fonts globally using the *Page Layout > Fonts* button. Alternatively you can select each chart element and alter the font from the *Home* menu.

Saving charts as Excel worksheets

If you want to save a chart in an Excel worksheet you need to move the chart to a separate worksheet, which you then save. You have several ways to do this:

- Make the chart in a new workbook right at the outset.
- Make the chart in a new worksheet. Then right-click the name tab of the worksheet and move it to a new workbook, which you then save.
- Move an existing chart to a new worksheet (use the *Design > Move Chart* button). Then move that to a new workbook (as in the previous option).

This saves a version of the chart that is linked to the original data. This means that you cannot edit the data unless the linked file can be found. It does allow you to alter the general appearance of the chart.

9.4 EXERCISES

1. The ____ tool can be helpful if you want to extract a portion of your dataset to share with someone else.
2. The most generally useable format for data is which of the following?

    A) Excel 97–2003 format (XLS).
    B) Portable data format (PDF).
    C) Comma separated values (CSV).
    D) Regular Excel format (XLSX).
    E) Open Document format (ODS).

3. Use Destination Styles (S) uses the most appropriate formatting from the target program (Word or PowerPoint). TRUE or FALSE?
4. Imagine you have made a bar chart with two grouping variables. Which of the following chart elements would you be most likely to omit?

    A) Legend.
    B) Title.
    C) Axis value labels.
    D) Variable units.
    E) Gridlines.

5. Look at the *women.xlsx* data. The data represents the height (inches) and weight (pounds) of American women aged 30–39. The data were collected sometime prior to 1975. Make a suitable chart for a report.

↘ Go to the website for support material.

Answers to these exercises can be found in Appendix 1.

9.5 SUMMARY

Topic	Key points
What you can share	You can share three main elements of your data: the data itself, numerical summaries (including pivot tables) and graphics (including sparklines).
Sharing the dataset	If you share the entire Excel file you include all the pivot tables and charts as well as any notes.
	Some older versions of Excel cannot read XLSX files. The XLS format cannot handle some elements (mostly confined to *SmartArt*).
	The *Advanced* filter can be used to filter a subset of your data to a new worksheet or workbook.
Different file formats	Most useful formats are Excel 97–2003 format (XLS) and comma delimited values (CSV).
	CSV files can only include one worksheet. CSV can be read by most computer programs, including Excel and word processors.
	Any number formatting in operation when data is saved as CSV is preserved.
Worksheets	You can save a single worksheet by right-clicking the name tab and making a copy to a new workbook.
	Multiple worksheets can be selected (use the Ctrl key with the mouse) and copied (or moved).
Paste options	The *Paste* button allows a range of options, so you can transfer data only without any formatting or linked to the original data.
	The *Paste Special* option allows *Paste* options and a few others.
	Word and PowerPoint have similar but subtly different *Paste* options.
	You can embed a spreadsheet into the target program, which can allow editing but can also substantially increase file size.
Sparklines	Sparklines can be transferred as graphics, which will not permit any further editing.
	If you transfer a pivot table or summary chart as a table you can add the sparklines to the target table, mimicking the layout of the spreadsheet.
	If you paste into a new program you can save the sparklines as a graphics file. PowerPoint is able to save in various graphics formats as well as PDF.
Graphs	Excel charts need additional formatting before you share them. Make sure axes are appropriately labelled (including units). The *Chart Tools* menus allow you to format and edit your charts.
	Avoid clutter on your charts; better to make two clear charts rather than one cluttered one.
	Excel charts can be copied to other programs. The *Paste* options allow you to transfer the chart as a graphic.
	If you paste into a new program you can save the chart as a graphics file. PowerPoint can save in various graphics formats as well as PDF.

Topic	Key points
PDF format	Excel can save cells, worksheets and charts as PDF files. In some older versions of Excel you may need to download an add-in.
	Some Office fonts (for example, Calibri and Cambria) are not recognized by PDF editing programs. It is best to use standard fonts. Use the *Page Layout* menu and change fonts globally or select each text element and alter the font using the *Home* menu.

APPENDIX 1

ANSWERS TO EXERCISES

The answers to the end-of-chapter exercises are set out in the following sections.

Chapter 1

1. FALSE. In scientific recording format you set out your data with columns for each variable. The rows are the individual records.
2. C. Individual items are called records, observations or replicates.
3. FALSE. The COUNTIF function is what you need to make index variables.
4. When you have your data in scientific recording format you can use your data like a **database**, and the key to accessing the information is the Excel **PivotTable** tool. A pivot table allows you to arrange and rearrange your data easily and quickly. You can use the *Report Filter* to select certain groups, which helps you to access your data like a database.
5. You need to click in the block of data then use the *Insert >PivotTable* button. Make sure the result is sent to a *New Worksheet*. Drag the fields from the list at the top of the *PivotTable Field List* task pane to the sections at the bottom: *Diet* to *Column Labels*, *Rep* to *Row Labels*, and *Size* to *Values*.

Chapter 2

1. B. If you use greater than symbols (or less than) you must enclose the entire criterion in quotes. A cell reference is fine; a cell reference in quotes is interpreted as text. A number can be in quotes or not.
2. TRUE. The DATEVALUE function takes a text string and converts it to an Excel date number.
3. When you need to select all the data in a worksheet you can use **Ctrl+A** as a shortcut from the keyboard rather than use the mouse. Holding down the control key and pressing the A key will select all the data in a block – most likely the entire worksheet.
4. Lookup tables can be useful to make **replacement** variables. The two main functions used are **VLOOKUP** and **HLOOKUP**.
5. You need to apply a filter. Click once in the block of data then use the *Data > Filter* button. Now click the triangle icon in the heading of the *Length* column to open the

filter dialog box. Click *Number Filters* then choose the *Above Average* option. You could also use the *Advanced* filter. You'd need to add a criterion for the mean value in a separate column. The formula for the criterion would need to be along these lines:
```
=IF(B2 > AVERAGE('Recording-layout'!$B$2 : $B$21), TRUE,
FALSE).
```

Chapter 3

1. All except D. Sorting can bring together similar items. Pivot tables arrange items in headings that are in order. Filter tools allow you to spot similar items and data validation tools allow you to restrict text entry. Chart tools (item D) are more suitable for numerical data.
2. FALSE. A pivot chart is suitable for some kinds of error checking (on grouped data and differences datasets), but scatter plots are especially suited to correlation and regression data. You cannot produce a scatter plot as a pivot chart.
3. The **dot** chart is a good tool for looking at numerical data in sample groups (usually differences data). You can adapt a **line** plot to make a dot chart by removing the lines.
4. E. All the other options are true. Validation on an already completed dataset is restricted to highlighting entries that do not match the entry criteria (the entries are circled in red).
5. You need to use the *Advanced* filter for this. You can filter in-place but you will need to make a new worksheet to hold the filter criteria. Copy the data headers to the new worksheet. Then add a column called *End* (column F). In cell F2 you'll need to use a formula that looks to see if the rightmost character is a space: `=EXACT(RIGHT(Original!A2, 1), " ")`. Return to the original data then use the *Data > Advanced* button. Select the *Filter the list, in-place* option. Select the data and criteria in the appropriate boxes. Once you click *OK* the filter is applied and you should see six records that match (that is, that have spaces at the end).

Chapter 4

1. To summarize a sample of data you need: an average (middle value), a measure of variability (dispersion), the sample size (replication) and an idea of the shape of the data (the distribution).
2. A histogram is a kind of **bar** chart that allows you to visualize the **distribution** of a data sample. The bars show the frequencies of the data in various size classes (bins).
3. FALSE. A line plot is useful for time-related data. The intervals on the x-axis are fixed and so the axis is categorical. Correlation and regression data require continuous axes, so a scatter plot is most useful for those kinds of dataset. A line plot can sometimes be used with correlation data but it is not the most useful type of chart.
4. B. You cannot make a scatter plot using a pivot chart. You can make a dot chart because it is simply a line plot without a line.
5. The *jackal.xlsx* file contains two worksheets: *Recording-layout* and *Sample-layout*.

You can use either to compute the appropriate values but the *Recording-layout* data needs to be processed through a pivot table to construct the samples. If you make a pivot table you'll need *Observations* in the *Row Labels* section, *Sex* in the *Column Labels* section and *Length* in the *Values* section. You can use the SKEW function for the skewness, the KURT function for kurtosis, AVERAGE for the mean and the MEDIAN function for the median. The results should look like Table A.1:

Table A.1 Shape statistics calculated for samples of mandible length from golden jackals.

	Male	Female
Skewness	0.09	-0.30
Kurtosis	0.18	-1.60
Mean	113.4	108.6
Median	113.5	109.0

The skewness is related to the difference between the mean and the median; if these values are close then skewness is low. Kurtosis is not directly related to the difference and is more of a pure shape statistic.

Chapter 5

1. B and D. There are 20 pairs of observations so degrees of freedom are 18. The critical value at the 5% level of significance is 0.444 (Table 5.1). You ignore the sign of the correlation coefficient so B and D exceed this, but A and C do not.
2. You can use **conditional formatting** to help you visualize the important relationships in a correlation matrix. You can set conditional formatting to highlight values that exceed the critical value for the sample size you have. You can also use it to apply a different format to negative correlations.
3. If you have a correlations or regressions dataset that includes grouping variables you can use a **pivot table** to help you produce a correlation matrix. A pivot table allows you to rearrange your data by sample group. This makes it easier to calculate correlation matrices.
4. TRUE. Residuals are calculated from the difference between actual values and predicted values (which is what the line of best fit represents). The larger the residuals the further from the line of best fit the data points lie.
5. The *women.xlsx* file contains two variables, *height* and *weight*. There may be some doubt about which variable should be considered the response and which the predictor. Assume that *weight* is the response. The easiest method is to use the *Analysis ToolPak*. Use the *Data > Data Analysis button* then choose *Regression*. Select the data in the appropriate sections. In the *Residuals* section tick the box labelled *Residuals*. Once the output is complete you can use the *Analysis ToolPak* again and choose *Histogram*. Select the *Residuals* data (if you used a new worksheet these values start around cell C25). Make sure you select *Chart Output*; your results will

form a histogram. You can also use the option *Normal Probability Plots* from the *Analysis ToolPak Regression* dialog to make a different sort of plot that can help you decide whether the residuals are normally distributed.

Chapter 6

1. FALSE. The general default is for gaps in the data to be shown as gaps in the sparklines but you can alter this to show a gap, zero or to interpolate.
2. On a line plot the intervals on the category axis are **equal**. The chart assumes the categories are all the same size (for example, same time interval).
3. C. You can use multiple fields. You can use the slicer at the same time as a report filter, field buttons or regular filters so items A, B and D are not strictly true. You can only use the slicer on a pivot table so E is not true.
4. FALSE. The field buttons allow you to access most of the filter tools as well as sorting options.
5. To begin you'll need to make a pivot table using *yr* as the *Row Labels*, *qtr* as the *Column Labels* and *gas* as the *Values*. Keep the grand totals, for the rows at least. The pivot table is slightly neater if you alter the layout to *Tabular*. At the bottom of the grand totals column, click in a blank cell and use *Insert > Line* in the *Sparklines* section. Highlight the column of totals. You now get an overview of the increasing consumption of gas over the period. Now highlight the cells to the right of the pivot table rows. Now use the *Line* button again to get sparklines for each year (miss out the year total and use only the quarterly values). Click on one of the sparklines and in the *Design* menu use the *Marker Color* button to highlight the high points. In nearly all cases the highest demand is the first quarter and the lowest the third.

Chapter 7

1. Association data is about frequencies of observations in **categories**. You may have data in one of three forms: **long records**, **short records** or a **contingency table**. Long records are in full scientific recording format, whilst short records have the frequencies calculated. The contingency table is the final summary table.
2. TRUE. Your data are frequencies of observations in various categories. So, you cannot use averages because you do not have replicated data.
3. FALSE. If you have a contingency table you can produce any kind of chart (within reason), so bar charts and column charts with or without stacked elements are perfectly possible. You can even make pie charts!
4. B. You can display *Series Name* but not series value.
5. You'll need to make a pivot table. There are many ways you can arrange the variables but the *Freq* field will have to be in the *Values* section. You can get a good overview by placing the fields as follows: *Survived* in *Column Labels*, *Sex* then *Age* in *Row Labels*, *Class* in *Report Filter*. You can then look at the different classes; third class doesn't appear to be so good for survivability of women and children. If you click *Options > PivotChart* you can visualize the data. A slicer tool can help to switch between the classes.

Chapter 8

1. C. Standard error (std. deviation ÷ √replicates) **cannot** be calculated directly by a pivot table. B (the mean) can be calculated using the *Average* option. E (replicates) can be calculated using the *Count* option.

2. A, B and F. Type B (dot chart) shows data in categories (that is, sample groups) so is a suitable type. Type C (line plot) is essentially a dot chart with the points joined but this implies a link between the categories, so is normally only used for time-related data. Type D (scatter plot) is used for correlations. Type E (pie chart) shows compositional data and is most used with association data.

3. FALSE. Error bars are a deviation and not absolute values. You can use the quartiles from the IQR but you need to work out the differences between the quartiles and the median as the values for the error bars.

4. Box-and-whisker plots usually show **non-parametric** statistics for example, the **median, IQR** and **range**. The median forms a point, the IQR forms the box part and the whiskers extend to the max–min (that is, the range).

5. You need to make a pivot table to help you rearrange the data and filter the year and species information. So, make a pivot table and arrange the fields as follows: *Qty* in the *Values* section, *Mng* in the *Column Labels* section, *Transect* and *Section* in the *Row Labels* section, *Spp* and *Year* in the *Report Filter*. You can now use the *Report Filter* to select 1997 and the *swf* species. Now you have two columns containing the abundance data. Blank cells are not missing; they represent a section where the management is one or the other, so for each row there is a value in one column or the other (but not both). You can now use the MEDIAN and QUARTILE functions to calculate the median and quartiles (first and third give you the IQR). The functions ignore blank cells (but not actual zero values). Your results should look like Table A.2:

Table A.2 Median and inter-quartiles for butterfly abundance samples in areas of management and no management (1997, species = swf).

	no	yes
Median	5.00	6.36
UQ	11.63	14.29
LQ	2.64	4.11

Now you have these results you can use the *Report Filter* to change the year or species and your results will update; however, some years have more observations so to leave room place your calculations further down. Make sure that your formulae include all the cells in the column above (remember, blank cells are ignored). If you accidentally overwrite the results table you can use the undo button and make a bit of room (you can split the screen so that you can always see the results).

Chapter 9

1. The **Advanced filter** tool can be helpful if you want to extract a portion of your dataset to share with someone else. You can send the filter results to a new location, which can be a new worksheet or workbook.
2. C. The CSV format is readable by most computer programs including Excel and Word. It is also readily imported by most statistical analysis programs.
3. TRUE. As the name suggests, the format of the destination is used as far as possible. For Word this results in a bare table. In PowerPoint the result is similar, a table with only basic formatting.
4. B. The title can be readily reproduced using a caption in a word processor. This leaves more room on the chart for what is most important, the data. Your caption can include more information than there is sensible room for on the chart.
5. The data are set out with a predictor variable, *height* and a response, *weight*. You'll need a scatter plot for this. The data happen to be in the appropriate Excel-ready format so click in the block of data then use the *Insert > Scatter* button and choose a basic plot. First of all you can delete the title, legend and gridlines. These can be selected simply by clicking in the chart but you can also select from a dropdown list on the *Layout* or *Format* menus. Once an item is selected, use the delete key on the keyboard. Now you need to change the axis scales: set the minimum to 55 and 110 for the *x* and *y* axes respectively. From the *Layout* menu add appropriate titles to the axes; don't forget the units. Go to the *Page Layout* dialog and alter the fonts to Arial for everything. You can now resize the chart; make it roughly square. The chart can now be copied to the clipboard or saved directly as a PDF. Your final chart should resemble Figure A.1.

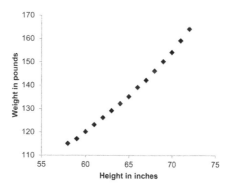

Figure A.1 The height (in) and weight (lbs) of a sample of American women from an unknown date before 1975.

APPENDIX 2

DIFFERENCES BETWEEN VERSIONS OF EXCEL

This book describes Excel 2010 for Windows in detail. At time of writing the latest version (2013) had recently been released. The three main versions of Excel that are likely to be in common use are 2007, 2010 and 2013. There are a few minor differences between these versions so if you are using 2007 or 2013 you can look here to see how to cope.

GENERAL DIFFERENCES

If you are using an older version of Excel then the differences are rather more pronounced and it would simply be too great a task to mention them all. However, the basic function set of Excel has changed very little over the years and the major charts and pivot tables are available on these versions (but not sparklines). Filter and sort tools and the *Analysis ToolPak* have also altered very little. The major difference is that older versions use the old-style menu system, rather than the ribbon. So, if you use an old version of Excel you should be able to manage but will need a little patience to find the appropriate menu buttons.

On a Mac Excel 2011 is broadly similar to Windows Excel 2010 but a few menu items are in a slightly different place. The *Analysis ToolPak* is not available. There are a few other minor differences but in the main you should be able to cope quite well.

If you use Open Office then you have similar issues to users of old Excel versions. In Open Office you do not have the ribbon but the main functions are the same. You can use pivot tables (which may be called the *Data Pilot*), although there are some limitations. Filters and sort tools are also available. The *Analysis ToolPak* is not available. The charting system in Open Office is more like the older versions of Excel and the main sorts of chart are available (but not sparklines). So, if you use Open Office you should be able to follow most of the ideas in this book and cope quite well.

In the notes that follow you'll see the main differences between the Windows versions 2010 and 2013 listed. Version 2007 is very similar to 2010 and the differences are so minor that they are only mentioned at the appropriate points.

Pivot tables

The main difference between the versions is with the names of the menus.

- 2010 – Options and Design.
- 2013 – Analyze and Design.

Essentially the *Options* menu has been renamed *Analyze*. All the tools that are in the *Options* menu in 2010 are available in the *Analyze* menu. There is a single additional button in 2013, *Insert Timeline*, which is a time-based slicer tool.

In 2007 you have no option to repeat item labels (*Design > Report Layout*). The slicer tool is not available. The only other difference is that calculated fields are inserted using *Options > Formulas > Calculated Field*.

Charts

Chart tools have not altered to any great extent; the main differences are in what the menus are called and the general appearance. The tools are all there, just accessed slightly differently. In 2013 one of the menus has been lost:

- 2010 – Design, Layout, Format and Analyze.
- 2013 – Analyze, Design and Format.

Most of the tools that were on the *Layout* menu in 2010 have been shifted into the *Design* menu: the *Add Chart Element* button brings down a list of items, which you can choose to add to the chart (Figure A.2).

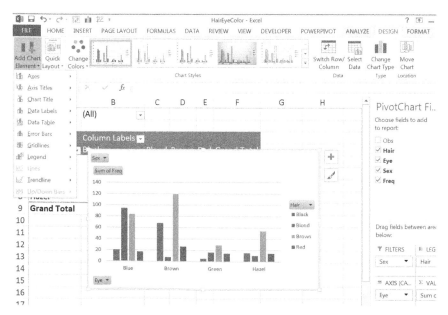

Figure A.2 The chart tools menus in Excel 2013 are slightly different to 2010.

The other main difference is that once you have a chart in Excel 2013 you can add elements using the + button that appears beside the chart (Figure A.2). The paintbrush icon allows you to change the overall style of the chart (not the type of chart). If you choose to format a chart element the formatting menus are slightly different to previous versions. Instead of the different formatting categories in a list on the left you get some icons across the top (Figure A.3).

Figure A.3 The formatting menu in Excel 2013 takes a more icon-led approach than earlier versions.

In Excel 2013 the *Insert* menu is still the place to insert a chart but the buttons have been reduced to icons. The main other difference is that you can temporarily remove a data series from a chart using the *Select Data* button, assuming that you have several series of course. You can also do this using the funnel icon that appears beside the chart (when you have multiple series).

Sparklines are not available in Excel 2007.

SUMMARY

In the main the differences between Excel versions 2007, 2010 and 2013 are minor. The only substantial feature to be added since 2007 is sparklines. Most of the other changes are either cosmetic or of no great importance to the topics covered in this book.

Excel is a powerful and useful tool and whichever version you use you should be able to unleash its potential to help you manage your data more effectively.

INDEX

CPSIA information can be obtained at www.ICGtesting.com
Printed in the USA
BVOW06*1247260116

434290BV00005B/12/P

9 781784 270087